Fundamentals of Data Analytics

Rudolf Mathar · Gholamreza Alirezaei ·
Emilio Balda · Arash Behboodi

Fundamentals of Data Analytics

With a View to Machine Learning

Springer

Rudolf Mathar
Institute for Theoretical Information Technology
RWTH Aachen University
Aachen, Nordrhein-Westfalen, Germany

Gholamreza Alirezaei
Chair and Institute for Communications Engineering
RWTH Aachen University
Aachen, Nordrhein-Westfalen, Germany

Emilio Balda
Institute for Theoretical Information Technology
RWTH Aachen University
Aachen, Nordrhein-Westfalen, Germany

Arash Behboodi
Institute for Theoretical Information Technology
RWTH Aachen University
Aachen, Nordrhein-Westfalen, Germany

ISBN 978-3-030-56833-7 ISBN 978-3-030-56831-3 (eBook)
https://doi.org/10.1007/978-3-030-56831-3

This Springer imprint is published by the registered company Springer Nature Switzerland AG
The registered company address is: Gewerbestrasse 11, 6330 Cham, Switzerland

Preface

Data Analytics is a fast developing interdisciplinary branch of science combining methods from exploratory statistics, algorithm and information theory to reveal structures in large data sets. Systematic patterns are often concealed by the high dimension and the sheer mass of the data. Diagram 1 visualizes which skills are important for successful data science. Computer Science, Statistics and substantive expertise in the respective application field contribute to the field of Data Science. Moreover, Machine Learning, statistical analysis and tailored applications all rely on methods from data analytics. To be a successful researcher in data science, one should be experienced in and open to methods from each of the domains.

The difference to classical approaches about 20 years ago mainly lies in the tremendous size of the data that can be handled and the huge dimension of observations. This was previously not accessible, but by the increasing efficiency and speed of computers and the development of parallel and distributed algorithms problems of unanticipated size can be solved at present.

Data analytics is a crucial tool for internet search, marketing, medical image analysis, production and business optimization and many others. Typical applications are, e.g., classification, pattern recognition, image processing, supervised and unsupervised learning, statistical learning and community detection in graph based data. By this, data analytics contributes fundamentally to the field of machine learning and artificial intelligence.

There is tremendous demand for corresponding methods. By digitization of industrial processes and the Internet of Things, huge amounts of data will be collected in short periods of time. It is of high value to analyze this data for the purpose of steering underlying production processes. Furthermore, social media and Internet searches generate huge amounts of diverse data, which have to be scrutinized for meaningful conclusions. Furthermore, healthcare and medical research rely on fast and efficient evaluation of data. It is estimated that in the near future more than one megabyte of data per person will be created each second.

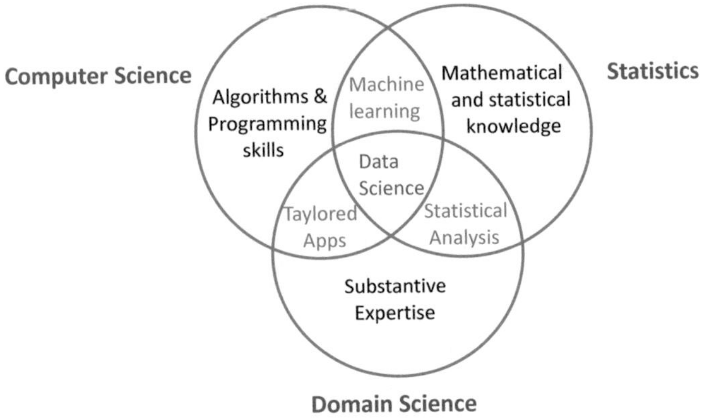

Fig. 1 Our view of locating the developing area of Data Science into established fields and where the domains contribute to data science

The main purpose of this book is to provide a comprehensive but concise overview of the fundamentals for analyzing big data. Much of the material needs a thorough understanding of mathematics. Readers should be familiar with basic probability theory, matrix algebra and calculus at an advanced undergraduate level. Any material exceeding this level will be discussed and proved in detail. Matrix algebra and matrix extremals are of particular interest in data analytics. A short introduction to basic theorems of matrix analysis and multivariate distributions is included. Optimization problems will play an important role throughout the book, so that an overview of convex programming and Lagrange duality theory is also presented. Emphasis is laid on clarity and completeness, such that students can develop a deep understanding of the principles throughout the book.

The book has erased from a series of lectures Rudolf Mathar has taught at RWTH Aachen University from the Winter term 2016 on. The lecture series was strongly supported by the co-authors Arash Behboodi and Emilio Balda, and tremendously enriched by exercise problems. They also developed large project exercises, which can only be solved by implementing corresponding algorithms, mainly TensorFlow and PyTorch libraries for Python. Gholamreza Alirezaei has taught the lecture at the Technical Universities of Munich, Dresden and Aachen. He has extended the material and contributed new proofs. The feedback from our students was overwhelmingly positive and we thank them for a number of valuable comments concerning structure and self-containment of the lecture.

Our answer to the question ‘Why writing another book on Data Analytics?’ is as follows. When designing the lecture, no textbook was found that presents the chosen material in a comprehensive, compact and self-contained way for graduate students in a three hours per week course over one semester. Moreover, our emphasis is to close the gap between mathematical foundations and algorithmic applications by carefully presenting the ground laying techniques. We also want to dig sufficiently deep into mathematical foundations and thus present the nontrivial background about matrix analysis and optimization. We think that broad knowledge of the underlying mathematics will enable students to engage in methodological research and successful applications. The emphasis of this text is not on numerical evaluation and programming paradigms, however, we briefly introduce corresponding tools and methods in Chapter 1.

We would like to thank Dr. Eva Hiripi from Springer, who after having a look at the preliminary material published on the institute web pages suggested to create an extended and refined version in the Springer Nature series. We appreciate her assistance during the whole publication process.

Aachen, Germany
July 2020

Rudolf Mathar
Gholamreza Alirezaei
Emilio Balda
Arash Behboodi

Contents

Acronyms

AI	Artificial Intelligence
ANN	Artificial Neural Networks
DFS	Distributed File Systems
GFS	Google file system
HDFS	Hadoop Distributed File System, Apache
MB	Megabyte
MC	Markov Chain
MDP	Markov Decision Process
MDS	Multi-dimensional Scaling
ML	Machine learning
MLE	Maximum Likelihood Estimator
NN	Neural Network
OMC	Optimum Margin Classifier
PCA	Principal Component Analysis
SMO	Sequential Minimal Optimization
SSQ	Sum of Squares
SVD	Singular Value decomposition
SVM	Support Vector Machine

Chapter 1
Introduction

Data Analytics is the science of exploring (big) data and designing methods and algorithms for detecting structures and information in the data. More specifically, we define Data Analytics as the discovery of models that capture the behavior of data and can be used to extract information, draw conclusions and make decisions. We conceive the concept of a "model" in a rather wide sense. For example, all of the following are regarded as a model which can be fitted to data.

- *Statistical models*: We assume that there is an underlying distribution that governs the occurrence of data. The objective is to specify this distribution, often by estimating corresponding parameters. We can also quantify how large the potential deviation from the true distribution may be. For example, given a set of real numbers we may assume that these numbers are independent samples from a Gaussian distribution and develop estimators of their expectation and variance. The model for data then is the Gaussian distribution $\mathcal{N}(\mu, \sigma^2)$, and the samples are stochastically independent realizations.
- *Machine learning*: A machine learning model uses the available data as a training set for fitting an algorithm. Typical representatives are Bayesian inference models, artificial neural networks, support-vector machines, decision trees, and many others. The trained model explains the data. These models are very common in practice, e.g., for pattern recognition, image processing, medical analysis and even for making cognitive decisions.
- *Dimensionality reduction*: The goal of these models is to extract the most prominent features of the data and ignore the rest. This could be in the form of separating noise from data to detect a relevant signal. For example, feature selection algorithms aim to select a subset of variables that better capture the relevant information contained in the data. Principal Component Analysis (PCA) projects high dimensional data onto a low dimensional space, so that they can be further processed. More sophisticated methods like diffusion maps and manifold learning algorithms

R. Mathar et al., *Fundamentals of Data Analytics*,
https://doi.org/10.1007/978-3-030-56831-3_1

make use of the similarities between data points to reveal underlying non-linear structures in the data.

- *Summarization models*: these methods are used to aggregate and summarize large amounts of data in a comprehensive ways. Clustering and discrimination methods are typical candidates. Samples that are close or near in a certain sense are combined as a cluster, which is represented by some cluster head.

Many aspects of machine learning are strongly intertwined with data analytics. Supervised learning algorithms are usually trained by data as is particularly the case for artificial neural networks or classifiers. Normally the amount of data is huge so that problems of efficient data processing are naturally inherent.

Without having a solid model for big data one may easily be deceived by certain patterns as is explained in the following example. In large random data sets, unusual features occur, which are purely the effect of randomness. This is called *Bonferroni's principle*.

Example 1.1 Imagine we want to find evil-doers by looking for people who both were in the same hotel on two different days, cf. [21, p. 6]. The general assumptions are as follows.

- There are 10^5 relevant hotels.
- Everyone goes to a hotel one day in one hundred.
- There are 10^9 people to be observed.
- People pick days and hotels independently at random.
- Examine hotel records for the last 1000 days.

The probability that any two people visit a hotel on any given day is equal to $\frac{1}{100} \cdot \frac{1}{100}$. The probability that they pick the same hotel is $\frac{1}{10^4} \cdot \frac{1}{10^5} = 10^{-9}$. Therefore, the probability that two people visit the same hotel on two different days are $10^{-9} \cdot 10^{-9} = 10^{-18}$.

The cardinality of the event space is the number of pairs of people $\binom{10^9}{2}$ multiplied by the pairs of days $\binom{10^3}{2}$. Hence, the expected number of events we are seeking for, using the approximation $\binom{n}{2} \approx \frac{n^2}{2}$, is given by

$$\binom{10^9}{2} \cdot \binom{10^3}{2} \cdot 10^{-18} \approx 5 \cdot 10^{17} \cdot 5 \cdot 10^5 \cdot 10^{-18} = 25 \cdot 10^4 = 250.000.$$

Hence, there are about 250.000 events which must be closely investigated if truly evil-doers have met in conspiracy.

We will now discuss some aspects of algorithms that are able to deal effectively with huge amounts of data. Corresponding computations may be very time and memory consuming, think, e.g., of multiplying matrices of dimension one million to one million and successively finding the largest eigenvalue. The main remedy to get results in short time is parallelism.

Fig. 1.1 Racks of compute nodes

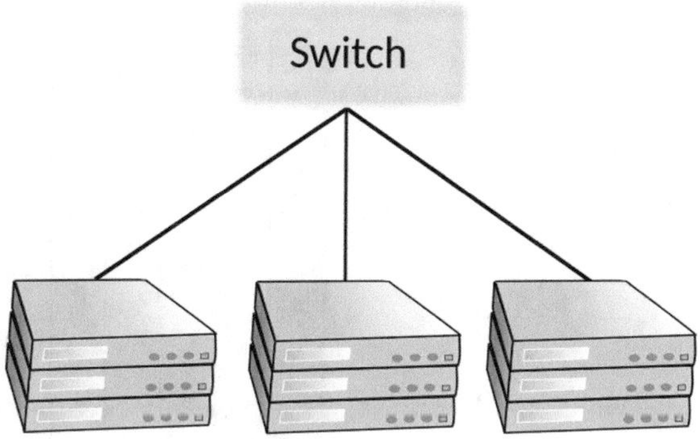

1.1 Parallel Programming and MapReduce

When the computation is to be performed on very large data sets, so called *big data*, it is not efficient to accommodate the whole data in a single database and perform the computations sequentially. The key idea is to use parallelism from "computing clusters" built of commodity hardware, instead of using a supercomputer. The clusters are connected by Ethernet or inexpensive switches.

The software stack consists of a Distributed File Systems (DFS) and MapReduce. In the distributed file system, files are divided into chunks (typically of 64 MB) and these chunks are replicated, typically 3 times on different racks. There exists a *file master node*, or *name node*, with information where to find copies of files. Some of the implementations of a DFS are GFS (Google file system), HDFS (Hadoop Distributed File System, Apache) and Cloud Store (open source DFS).

On the other hand, MapReduce is the computing paradigm. With MapReduce, the system manages parallel execution and coordination of tasks. Two functions are written by users, namely Map and Reduce. The advantage of this system is its robustness against hardware failures and its ability to handle big data. MapReduce has been implemented internally by Google, it is used for carrying out computations on their servers.

The architecture of this system is such that compute nodes are stored on racks, each with its own processor and storage device. The racks are connected by fast Gigabit links, as presented in Fig. 1.1. The principles of this system are as follows. First, files are stored redundantly to protect against failure of nodes. Second, computations are divided into independent tasks. If one fails it can be restored without affecting others.

1.2 MapReduce for Linear Algebra

We discuss an example of implementation of matrix-vector multiplication using MapReduce [21]. For that, suppose that the matrix $\mathbf{M} \in \mathbb{R}^{m\times n}$ and the vector $\mathbf{v} \in \mathbb{R}^n$ are given, and the goal is to compute their multiplication $\mathbf{x} = \mathbf{Mv}$ (Fig. 1.2):

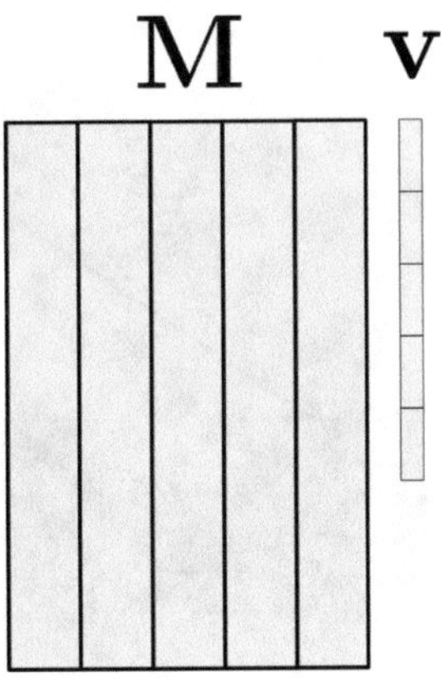

Fig. 1.2 Matrix-Vector Multiplication

$$x_i = \sum_{j=1}^{n} m_{ij} v_j.$$

Example 1.2 (*Matrix-Vector Multiplication by MapReduce*) When n is large, say 10^7, then the direct computation requires the storage of the whole matrix which might not be efficient. Particularly in practice the matrix **M** can be sparse with say 10 or 15 non-zeros per row.

First the matrix is stored as pairs (i, j, m_{ij}), and the vector is stored as (i, v_i). MapReduce consists of two main functions, Map function and Reduce function. To implement the multiplication using MapReduce, the Map function produces a key-value pair for each entries of the matrix and the vector. To the entry m_{ij} the pair $(i, m_{ij}v_j)$ is associated where i is the key and $m_{ij}v_j$ is the pair. Note that it is assumed here that m is small enough to store the vector **v** in its entirety in the memory. The Reduce function receives all the key-value pairs, lists all pairs with key i and sums their values to get $(i, \sum_{j=1}^{n} m_{ij}x_j)$, which gives the i-th entry of the product.

If the vector **v** cannot be fitted into the memory, then the matrix **M** is divided into horizontal stripes of a certain width and the vector **v** is divided into vertical stripes of the same size as the matrix stripes' width. Accordingly, the multiplication can be divided into sub-tasks, each feasible using MapReduce.

This concept can be extended to matrix-matrix multiplication as is demonstrated in the following example.

Example 1.3 (*Matrix-Matrix Multiplication by MapReduce*) Given two matrices $\mathbf{M} \in \mathbb{R}^{n\times m}$ and $\mathbf{N} \in \mathbb{R}^{m\times r}$, the goal is to compute **MN**. Map function generates the following key-value pairs:

- For each element m_{ij} of **M** produce r key-value pairs $\big((i,k), (\mathbf{M}, j, m_{ij})\big)$ for $k = 1, \ldots, r$.

- For each element n_{jk} of $\mathbf{N}$ produce n key-value pairs $\big((i, k), (\mathbf{N}, j, n_{jk})\big)$ for $i = 1, \ldots, n$.

The Reduce function computes the multiplication as follows:

- For each key (i, k), find the values with the same j.
- Multiply m_{ij} and n_{jk} to get $m_{ij}n_{jk}$.
- Sum up all $m_{ij}n_{jk}$ over j to get $\sum_{j=1}^{m} m_{ij}n_{jk}$.

1.3 MapReduce for Relational Algebra

As we saw in the previous section, MapReduce can be used to perform various algebraic operations. As we will see throughout this book, we frequently need to search over the database and select or group data samples for further analysis. These operations are relational. This means that we will investigate data points based on the relation they have with others.

A relation R is defined by a sequence of attributes $A_1, \ldots, A_k$ and is denoted by $R(A_1, \ldots, A_k)$. The set of attributes is called the *schema* of the relation. For example, in a directed graph, the relation of "being connected" consists of two attributes. The first attribute is the starting vertex, and the second one is the target vertex that is connected to the first vertex via an incoming edge. Tuples in relation are connected vertices. The schema of this relation consists of two vertices.

Relational algebra is defined by some standard operations that are applied to relations. These operations are, among others, selection, projection, union, intersection, different, natural join, grouping and aggregation. We discuss some of these operations in detail and discuss how MapReduce can be used to implement them.

We first consider first the operation "selection". For a given relation R and a condition C, a selection operation, as the name suggests, returns only those tuples in R that satisfy the condition C. The selection operation is denoted by $\sigma_C(R)$. For example, finding the neighbors closer than r in a geometric graph is a selection operation. The relation R is the neighborhood relation. The constraint C on a given tuple is satisfied if their distance is smaller than r. For large datasets, this operation can be implemented using MapReduce. The Map and Reduce functions are defined as follows.

- **Map function:** for each tuple t in R, if t satisfies C, generate a key-value pair (t, t).
- **Reduce function:** for each pair (t, t), return the same key-value pair (t, t).

Hence, the selection operation is fully completed in the Map phase.

Consider the example of a directed graph. Suppose that we would like to select those vertices with at least one outgoing edge. Remember that the relation R defined on a directed graph has two attributes, the starting vertex and the target vertex. To find the vertices with at least one outgoing edge, it suffices to select the first attribute from the relation R. This is done via the projection operation. The projection operation acts on the relation R by selecting only a subset S of the attributes from each tuple in the relation. The output is denoted by $\pi_S(R)$. For large datasets, the MapReduce operation is implemented as follows:

- **Map function:** for each tuple t in R, return only those attributes in S as t', and generate key-value pair (t', t').
- **Reduce function:** for all pairs (t', t'), return a single key-value pair (t', t').

Note that in the map phase, there may be repeated key-value pairs. Therefore, in the Reduce function, only one of them is returned.

Grouping and aggregation are two other operations on relations. They arise often in machine learning algorithms. Consider again a directed graph. An example of the grouping operation is to create a list of neighbors of each vertex. In general, the grouping operation partitions the tuples in relation R based on common attributes in a subset of all attributes G. The aggregation operation acts on the groups by reducing the group to a single value. In the above example, the aggregation operation can be simply counting the number of neighbors. MapReduce for this grouping-aggregation operation consists, quite naturally, of a Map phase for grouping and a Reduce phase for aggregation. For a relation R, assume that the attributes in G are chosen for the grouping operation, and the aggregation function is given by $\gamma(\cdot)$. The MapReduce phase then consists of:

- **Map function:** for each tuple g with attributes in G, and for each tuple t in R, return those attributes in G^c as t', and generate key-value pair (g, t').
- **Reduce function:** for all pairs (g, t'), return a single key-value pair $(g, \gamma(t'))$.

Another common operation on relations is natural join. In the example of a directed graph, the tuples (u, v) in the relation R are connected by an edge outgoing from u and incoming to v. We define a new relation to specify all paths of length two in the graph. The relation has three attributes: start vertex, intermediate vertex and target vertex. Each tuple in relation can be represented by (u, v, w). We can, however, represent the new relation using the previous relation R, because each path of length two, say (u, v, w), consists of two pairs of connected vertices, (u, v) and (v, w) that share the vertex v. The new relation is the natural join of R by itself. More generally, the natural join of two relations R and S is denoted by $R \bowtie S$. It is constructed from tuples in R and S that agree on the common attributes of R and S. The common attributes are kept in the final schema and joined by the other attributes of tuples. Here, the new relation is the natural join of $R(U, V)$ and $R(V, W)$.

MapReduce for the natural join $R \bowtie S$ of $R(A, B)$ and $S(B, C)$ is given by:

- **Map function:** for each tuple (a, b) in R, generate the key-value pair $(b, (R, a))$. Similarly for S, generate for each tuple (b, c) the key-value pair $(b, (S, c))$.
- **Reduce function:** for all key-value pairs, with key b, take all a from values (R, a) and all c from values (S, c). Return all pairs (a, b, c).

Since the relations are defined over sets, we can expect to run set operations on relations as well. Set operations are applied to two relations R_1 and R_2. For example, the union of R_1 and R_2 consists of all tuples that are both in relation

R_1 and R_2. Therefore, the relations must share the same schema for set operations to be well defined. All set operations can be performed by combining complement and union. However, the complement operation can be expensive especially for large data sets. Therefore, in practice we will add more operations that are computationally less extensive. In general, union, intersection and difference are set operations with ready to use implementations. MapReduce for these operations are straightforward (see Exercise 1.3).

We conclude this chapter by considering an example of using MapReduce for some selected operations. Consider n data points $x_1, x_2, \ldots, x_n$ in a Euclidean space. We want to identify the vectors which are closer to one of two given points μ_1 and μ_2, and group them accordingly. This is a selection operation. Thereafter, we want to update μ_1 and μ_2 by the mean value of points in each group, which constitutes an aggregation operation. Similar to previous examples, we assume that n is large so that MapReduce is required for our operations. The Map part generates key-value pairs $(i : (x_j, 1))$ where i is equal to one if $\|x_j - \mu_1\|_2 < \|x_j - \mu_2\|_2$ and equal to two otherwise. At the end of the Map phase, all pairs with key i contain values with points closest to μ_i. The Reduce phase simply consists of adding all the values with the same key. This yields $(\sum_{x_j \in S_i} x_j, |S_i|)$ where S_i is the set of closest points to μ_i. The point μ_i is then updated as

$$\mu_i \rightarrow \frac{\sum_{x_j \in S_i} x_j}{|S_i|}.$$

As we will see later, these operations are crucial for the k-means clustering algorithm.

As a good exercise for dealing with big data, we recommend to design MapReduce implementations of the algorithms throughout this book for large data sets.

1.4 Exercises

Exercise 1.1 *(Matrix Vector Multiplication by Map-Reduce)*
Let $\mathbf{A} \in \mathbb{R}^{n\times m}$ be a matrix of large dimensions n and m.

(a) Let $\mathbf{v} \in \mathbb{R}^m$ be a vector. Describe an implementation of the multiplication of $\mathbf{A}$ and $\mathbf{v}$ using MapReduce.
(b) Let $\mathbf{B} \in \mathbb{R}^{m\times k}$ be a matrix. Describe how the multiplication of $\mathbf{A}$ and $\mathbf{B}$ can be performed by MapReduce.

Exercise 1.2 *(Sparse Vectors with MapReduce)*
Let $\mathbf{v} \in \mathbb{R}^n$ be a sparse vector of large dimension n.

(a) Let $\mathbf{w} \in \mathbb{R}^n$ be a sparse vector. How can the sum of $\mathbf{v}$ and $\mathbf{w}$ be carried out by MapReduce.
(b) Find an implementation of computing the average squared value $\frac{1}{n}\sum_{i=1}^{n} v_i^2$ by MapReduce.

Exercise 1.3 Specify the MapReduce operations for union, intersection and complement. How can the intersection of sets be implemented by MapReduce using union and complement?

Exercise 1.4 Consider a directed graph. A triple (u, v, w) of vertices is an element of relation S if there is directed cycle of length three through these vertices. Represent S as the natural join of the relation R, defined between vertices if they are connected.

Chapter 2
Prerequisites from Matrix Analysis

Linear algebra and matrix algebra provide the methodology for mapping high-dimensional data onto low-dimensional spaces. The combination of matrix analysis and optimization theory is of particular interest. This chapter focuses on elaborating tools which are prerequisite for data analytics and data processing. We will not only provide a vast overview, but will also introduce relevant theorems in detail with the derivation of proofs. We think that having deep insight into the general mathematical structure of matrix functions is extremely useful for dealing with unknown future problems.

After fixing the notation, the decomposition of matrices and the eigenvalue problem, matrix norms and partitioned matrices are discussed. Matrix ordering and matrix monotone functions are introduced to generalize the concept of monotonicity to the space of symmetric matrices. Further, stochastic and permutation matrices are used to solve matrix optimization problems. Finally, we introduce matrix approximations, isometry and projections.

Throughout the book the set of natural, integer, real, and complex numbers are denoted by $\mathbb{N}$, $\mathbb{Z}$, $\mathbb{R}$, and $\mathbb{C}$, respectively, while $\mathbb{R}_+$ indicates the set of nonnegative reals. The sets (a, b), $(a, b]$, and $[a, b]$ denote open, half-open, and closed intervals. Other sets are normally written by calligraphic letters. The union, the intersection, and the set theoretic difference of $\mathcal{A}$ and $\mathcal{B}$ are denoted by $\mathcal{A} \cup \mathcal{B}$, $\mathcal{A} \cap \mathcal{B}$, and $\mathcal{A} \setminus \mathcal{B}$, respectively.

The optimal value of an optimization variable x is highlighted by a superscript star as x^*. We write λ^+ for the positive part of a real number λ, i.e., $\lambda^+ = \max\{0, \lambda\}$.

Vectors are denoted by boldface lowercase letters. $\mathbf{0}_n$ and $\mathbf{1}_n$ are the all-zero and all-one vector of dimension n, respectively. The canonical basis vectors of $\mathbb{R}^m$ are written as $\mathbf{e}_1, \dots, \mathbf{e}_m$. The Euclidean norm of $\mathbf{x} \in \mathbb{R}^m$ is denoted by $\|\mathbf{x}\|$ or $\|\mathbf{x}\|_2$. Boldface uppercase characters indicate matrices. A matrix $\mathbf{A}$ of size $m \times n$ with entries a_{ij} is written as

$$\mathbf{A} = \mathbf{A}_{m \times n} = \left(a_{ij}\right)_{\substack{1 \le i \le m \\ 1 \le j \le n}}.$$

R. Mathar et al., *Fundamentals of Data Analytics*,
https://doi.org/10.1007/978-3-030-56831-3_2

$\mathbf{A}^T$ and $\mathbf{A}^{-1}$ are the transpose and the inverse of some matrix $\mathbf{A}$. The determinant of $\mathbf{A}$ is denoted by $\det(\mathbf{A})$, alternatively also by $|\mathbf{A}|$.

Some special matrices are the all-zero matrix $\mathbf{0}_{m\times n}$, the all-one matrix $\mathbf{1}_{m\times n}$, and the identity matrix $\mathbf{I}_n$. Diagonal matrices with all nondiagonal entries zero are denoted by $\mathbf{\Lambda} = \operatorname{diag}(\lambda_1, \lambda_2, \dots, \lambda_m)$. A matrix $\mathbf{U} \in \mathbb{R}^{m\times m}$ is called orthogonal (or sometimes orthonormal) if $\mathbf{U}\mathbf{U}^T = \mathbf{U}^T\mathbf{U} = \mathbf{I}_m$ holds. We denote the set of orthogonal matrices of size $m \times m$ by $\mathfrak{O}_m$, i.e.,

$$\mathfrak{O}_m = \left\{\mathbf{U} \in \mathbb{R}^{m\times m} \mid \mathbf{U}\mathbf{U}^T = \mathbf{U}^T\mathbf{U} = \mathbf{I}_m\right\}.$$

The *image* or *column space* of some matrix $\mathbf{M} \in \mathbb{R}^{m\times n}$ is defined as

$$\operatorname{Img}(\mathbf{M}) = \left\{\mathbf{M}\mathbf{x} \mid \mathbf{x} \in \mathbb{R}^n\right\},$$

the *kernel* or *nullspace* is defined by

$$\operatorname{Ker}(\mathbf{M}) = \left\{\mathbf{x} \in \mathbb{R}^n \mid \mathbf{M}\mathbf{x} = \mathbf{0}_m\right\}.$$

Furthermore, the *orthogonal complement* of $\mathcal{V} \subset \mathbb{R}^n$ is defined by

$$\mathcal{V}^\perp = \left\{\mathbf{x} \in \mathbb{R}^n \mid \mathbf{y}^T\mathbf{x} = 0 \text{ for all } \mathbf{y} \in \mathcal{V}\right\}.$$

Hence, any two vectors $\mathbf{y} \in \mathcal{V}$ and $\mathbf{x} \in \mathcal{V}^\perp$ are orthogonal. The *linear span* or *linear hull* of vectors $\mathbf{v}_1, \mathbf{v}_2, \dots, \mathbf{v}_n \in \mathbb{R}^m$ is the linear subspace formed of all linear combinations

$$\operatorname{Span}\big(\{\mathbf{v}_1, \mathbf{v}_2, \dots, \mathbf{v}_n\}\big) = \left\{\sum_{i=1}^n \alpha_i \mathbf{v}_i \mid \alpha_1, \alpha_2, \dots, \alpha_n \in \mathbb{R}\right\} \subseteq \mathbb{R}^m.$$

The spanning vectors may be linearly dependent, so the dimension of the linear span is at most $\min\{m, n\}$.

2.1 Decomposition of Matrices and Eigenvalues

For a square matrix $\mathbf{M} \in \mathbb{R}^{m\times m}$ consider the equation $\mathbf{M}\mathbf{v} = \lambda\mathbf{v}$ with $\lambda \in \mathbb{R}$ and $\mathbf{v} \in \mathbb{R}^m$, $\mathbf{v} \neq \mathbf{0}$. Geometrically this means that under transformation by $\mathbf{M}$ vector $\mathbf{v}$ experiences only a change in length or sign. The direction of $\mathbf{M}\mathbf{v}$ is the same as that of $\mathbf{v}$, so that $\mathbf{v}$ is stretched or shrunk or flipped. Vectors with this property are called *eigenvectors*, the scaling factor λ is called *eigenvalue*. It is clear that with any eigenvector $\mathbf{v}$ all multiples are also eigenvectors. This is the reason why we mostly consider eigenvectors to be normalized to length one. In this section we will deal with the problem of finding eigenvalues and eigenvectors of a given matrix $\mathbf{M}$.

Obviously the equation $\mathbf{Mv} = \lambda\mathbf{v}$ is equivalent to the so called *eigenvalue equation* $(\mathbf{M} - \lambda\mathbf{I})\mathbf{v} = \mathbf{0}$. This is equivalent to finding some $\boldsymbol{\lambda}$ such that $\det(\mathbf{M} - \lambda\mathbf{I}) = \mathbf{0}$. We will see that solution pairs $(\lambda, \mathbf{v})$ always exist, if the matrix $\mathbf{M}$ is symmetric.

As a generalization, we will also consider the so-called *singular value equation* $\mathbf{Mw} - \sigma\mathbf{u} = \mathbf{0}$ with potentially rectangular $\mathbf{M}$, and find solutions $(\sigma, \mathbf{u}, \mathbf{w})$. The eigenvalue and singular value equations will be shown to be closely related. Finally, we will present a numerical algorithm which converges to a solution of the eigenvalue equation.

Theorem 2.1 (Spectral Decomposition) *For a symmetric matrix* $\mathbf{M} \in \mathbb{R}^{n\times n}$ *there exist an orthogonal matrix* $\mathbf{V} \in \mathfrak{O}_n$ *and a diagonal matrix* $\boldsymbol{\Lambda} = \mathrm{diag}(\lambda_1, \ldots, \lambda_n) \in \mathbb{R}^{n\times n}$ *such that*

$$\mathbf{M} = \mathbf{V}\boldsymbol{\Lambda}\mathbf{V}^T . \tag{2.1}$$

The diagonal elements λ_i *of* $\boldsymbol{\Lambda}$ *are called* eigenvalues *of* $\mathbf{M}$, *the columns* $\mathbf{v}_i$ *of* $\mathbf{V}$ *are called the* eigenvectors *of* $\mathbf{M}$, *satisfying*

$$\mathbf{Mv}_i = \lambda_i\mathbf{v}_i$$

for all $i = 1, 2, \ldots, n$.

The right hand side of representation (2.1) is called *spectral decomposition* of $\mathbf{M}$. A proof of Theorem 2.1 can be found, e.g., in [19, p. 134].

Remark 2.2 Some number λ is an eigenvalue of the square matrix $\mathbf{M}$ if $\det(\mathbf{M} - \lambda\mathbf{I}) = 0$. Zero determinant means that $\mathbf{M} - \lambda\mathbf{I}$ is singular. Hence there exists some vector $\mathbf{v} \neq \mathbf{0}$ with $(\mathbf{M} - \lambda\mathbf{I})\mathbf{v} = \mathbf{0}$.

Remark 2.3 From Eq. (2.1) it follows that a square matrix $\mathbf{M}$ can be written as a superposition of rank-one matrices in the form

$$\mathbf{M} = \sum_{i=1}^{n} \lambda_i \mathbf{v}_i \mathbf{v}_i^T . \tag{2.2}$$

The set of eigenvalues defined by

$$\left\{\lambda \in \mathbb{C} \mid \det(\mathbf{M} - \lambda\mathbf{I}) = 0\right\}$$

is called the *spectrum* of $\mathbf{M}$.

By Theorem 2.1 all eigenvalues of a symmetric matrix are real. Hence, we can define the concept of *definiteness* for symmetric matrices on the basis of their eigenvalues.

Lemma 2.4 *Let* $\mathbf{M} \in \mathbb{R}^{n\times n}$ *be a symmetric matrix with eigenvalues* $\lambda_1, \ldots, \lambda_n$.

(a) $\mathbf{M}$ *is called* nonnegative definite *(*positive definite*), if* $\lambda_i \geq 0$ *(*$\lambda_i > 0$*) for all* $i = 1, 2, \ldots, n$.

(b) *If* $\mathbf{M}$ *is nonnegative definite, there exists a so-called* Cholesky decomposition

$$\mathbf{M} = \mathbf{V}\mathbf{\Lambda}\mathbf{V}^T = \mathbf{V}\mathbf{\Lambda}^{1/2}\big(\mathbf{V}\mathbf{\Lambda}^{1/2}\big)^T = \mathbf{C}\mathbf{C}^T \tag{2.3}$$

with $\mathbf{\Lambda}^{1/2} = \operatorname{diag}\big(\lambda_1^{1/2}, \ldots, \lambda_n^{1/2}\big)$.

(c) *If* $\mathbf{M}$ *is nonnegative definite, then* $\mathbf{x}^T\mathbf{M}\mathbf{x} = \mathbf{x}^T\mathbf{C}\mathbf{C}^T\mathbf{x} \geq 0$ *holds for all* $\mathbf{x} \in \mathbb{R}^n$.

(d) *If* $\mathbf{M}$ *is positive definite,* $\mathbf{x}^T\mathbf{M}\mathbf{x} = \mathbf{x}^T\mathbf{V}\mathbf{\Lambda}\mathbf{V}^T\mathbf{x} = \mathbf{y}^T\mathbf{\Lambda}\mathbf{y} = \sum_{i=1}^n \lambda_i y_i^2 > 0$ *holds for all* $\mathbf{x} \in \mathbb{R}^n \setminus \{\mathbf{0}\}$.

Obviously, the identity matrix $\mathbf{I}_n$ is positive definite. The system of canonical unit vectors $\mathbf{e}_i$, $i = 1, 2, \ldots, n$, forms a corresponding system of orthonormal eigenvectors, each with eigenvalue one. The columns of any other orthogonal matrix can also serve as a system of eigenvectors to eigenvalue 1.

Example 2.5 Let $\mathbf{M} = \mathbf{A} + \mathbf{I}_n \in \mathbb{R}^{n\times n}$ be a real symmetric matrix and $\mu_1, \ldots, \mu_n$ be the eigenvalues of $\mathbf{A}$ with corresponding orthonormal eigenvectors $\mathbf{v}_1, \ldots, \mathbf{v}_n$. It holds that

$$\mathbf{M}\mathbf{v}_i = \mathbf{A}\mathbf{v}_i + \mathbf{v}_i = \mu_i\mathbf{v}_i + \mathbf{v}_i = (\mu_i + 1)\mathbf{v}_i,\ i = 1, \ldots, n,$$

such that $\mathbf{M}$ and $\mathbf{A}$ have the same eigenvectors. The eigenvalues of $\mathbf{M} = \mathbf{A} + \mathbf{I}_n$ are $\lambda_i = \mu_i + 1$ for all $i = 1, \ldots, n$. Hence, if $\mathbf{A}$ is nonnegative definite, then $\mathbf{M}$ is positive definite.

Example 2.6 Let $k \in \mathbb{N}$ and $\mathbf{M} \in \mathbb{R}^{n\times n}$ be a symmetric matrix with eigenvalues $\lambda_1, \ldots, \lambda_n$ and eigenvectors $\mathbf{v}_1, \ldots, \mathbf{v}_n$. Then by iterating $\mathbf{M}^k\mathbf{v}_i = \mathbf{M}^{k-1}\mathbf{M}\mathbf{v}_i = \mathbf{M}^{k-1}\lambda_i\mathbf{v}_i$ we obtain

$$\mathbf{M}^k\mathbf{v}_i = \lambda_i^k\mathbf{v}_i,\ i = 1, 2, \ldots, n\,. \tag{2.4}$$

This shows that $\mathbf{M}^k$ has the same eigenvectors as $\mathbf{M}$ with corresponding eigenvalues λ_i^k, $i = 1, \ldots, n$. Hence, $\mathbf{M}^k$ can be represented as

$$\mathbf{M}^k = \sum_{i=1}^n \lambda_i^k\mathbf{v}_i\mathbf{v}_i^T\,. \tag{2.5}$$

If k is even, then $\mathbf{M}^k$ is nonnegative definite, otherwise $\mathbf{M}^k$ and $\mathbf{M}$ have the same number of negative, positive and zero eigenvalues.

To decompose arbitrary, even rectangular matrices we next consider the *singular value decomposition*, which can be seen as a generalization of the spectral decomposition.

Theorem 2.7 (Singular Value Decomposition) *For each* $\mathbf{M} \in \mathbb{R}^{m\times n}$ *there exist orthogonal matrices* $\mathbf{U} \in \mathcal{O}_m$ *and* $\mathbf{W} \in \mathcal{O}_n$*, and* $\mathbf{\Sigma} \in \mathbb{R}^{m\times n}$ *with nonnegative entries on its diagonal and zeros otherwise such that*

$$\mathbf{M} = \mathbf{U}\,\mathbf{\Sigma}\,\mathbf{W}^T\,. \tag{2.6}$$

The diagonal elements of $\mathbf{\Sigma}$ *are called* singular values *of* $\mathbf{M}$*, the columns of* $\mathbf{U}$ *and* $\mathbf{W}$ *are called the* left *and the* right singular vectors *of* $\mathbf{M}$*, respectively.*

A proof of Theorem 2.7 can be found in [30, p. 411].

Remark 2.8 Let $\sigma_1, \sigma_2, \ldots, \sigma_{\min\{m,n\}}$ denote the diagonal entries of $\mathbf{\Sigma}$ and $\mathbf{U} = (\mathbf{u}_1, \mathbf{u}_2, \ldots, \mathbf{u}_m)$ and $\mathbf{W} = (\mathbf{w}_1, \mathbf{w}_2, \ldots, \mathbf{w}_n)$. For all $i = 1, 2, \ldots, \min\{m, n\}$ it holds that

$$\mathbf{u}_i^T\mathbf{M} = \sigma_i\mathbf{w}_i^T \quad \text{and} \quad \mathbf{M}\mathbf{w}_i = \sigma_i\mathbf{u}_i\,. \tag{2.7}$$

If $m \neq n$ then for all $\min\{m, n\} < i \leq \max\{m, n\}$ both right hand sides in (2.7) are equal to $\mathbf{0}$. Similar to (2.2) we obtain

$$\mathbf{M} = \sum_{i=1}^{\min\{m,n\}} \sigma_i\mathbf{u}_i\mathbf{w}_i^T\,.$$

Both the singular value decomposition and the spectral decomposition are closely related as follows. Let $\mathbf{M} \in \mathbb{R}^{m\times n}$ and $k = \min\{m, n\}$. Using the singular value decomposition $\mathbf{M} = \mathbf{U}\mathbf{\Sigma}\mathbf{W}^T$ with orthogonal matrices $\mathbf{U}$ and $\mathbf{W}$ we obtain

$$\mathbf{M}\mathbf{M}^T = \mathbf{U}\mathbf{\Sigma}\mathbf{W}^T\mathbf{W}\mathbf{\Sigma}^T\mathbf{U}^T = \mathbf{U}\,\mathrm{diag}\underbrace{(\sigma_1^2, \ldots, \sigma_k^2, 0, \ldots, 0)}_{m \text{ entries}}\mathbf{U}^T$$

and

$$\mathbf{M}^T\mathbf{M} = \mathbf{W}\mathbf{\Sigma}^T\mathbf{U}^T\mathbf{U}\mathbf{\Sigma}\mathbf{W}^T = \mathbf{W}\,\mathrm{diag}\underbrace{(\sigma_1^2, \ldots, \sigma_k^2, 0, \ldots, 0)}_{n \text{ entries}}\mathbf{W}^T\,,$$

which shows that $\mathbf{U}$ and $\mathbf{W}$ are the eigenvector matrices of $\mathbf{M}\mathbf{M}^T$ and $\mathbf{M}^T\mathbf{M}$, respectively. Furthermore, the singular values of $\mathbf{M}$ coincide with the square roots of the eigenvalues of $\mathbf{M}\mathbf{M}^T$ or $\mathbf{M}^T\mathbf{M}$, which are nonnegative due to $\mathbf{x}^T\mathbf{M}^T\mathbf{M}\mathbf{x} = \mathbf{y}^T\mathbf{y} \geq 0$. Hence, by these equations one can calculate the singular value decomposition of any matrix $\mathbf{M}$ by the spectral decomposition of the products $\mathbf{M}\mathbf{M}^T$ and $\mathbf{M}^T\mathbf{M}$.

In most data science applications only a few dominant eigenvalues with corresponding eigenvectors are needed. For this purpose there exist iterative methods, one of which is the so-called *von Mises iteration* [31] or the *power iteration*, which determines solely the largest eigenvalue and the corresponding eigenvector. Von Mises iteration is a very simple algorithm, which may converge slowly.

Theorem 2.9 (R. von Mises and H. Pollaczek-Geiringer, 1929) *Let* $\mathbf{M} \in \mathbb{R}^{n\times n}$ *be symmetric with eigenvalues* $|\lambda_1| > |\lambda_2| \geq \cdots \geq |\lambda_n|$ *and corresponding orthonormal eigenvectors* $\mathbf{v}_1, \mathbf{v}_2, \ldots, \mathbf{v}_n$. *Assume that the initial vector* $\mathbf{y}^{(0)}$ *is not orthogonal to the dominant eigenvector* $\mathbf{v}_1$. *Then the following iteration over* $k \in \mathbb{N}$

$$\mathbf{y}^{(k)} = \mathbf{M}\,\mathbf{y}^{(k-1)}, \quad \mathbf{x}^{(k)} = \frac{\mathbf{y}^{(k)}}{\|\mathbf{y}^{(k)}\|}$$

and

$$\mu^{(k)} = \frac{\left(\mathbf{y}^{(k-1)}\right)^T \mathbf{y}^{(k)}}{\|\mathbf{y}^{(k-1)}\|^2}$$

yields

$$\left|\mathbf{v}_1^T \lim_{k\to\infty} \mathbf{x}^{(k)}\right| = 1$$

and

$$\lim_{k\to\infty} \mu^{(k)} = \lambda_1 \,.$$

Hence, $\mathbf{x}^{(k)}$ *converges to the direction of the dominant eigenvector* $\mathbf{v}_1$ *and* $\mu^{(k)}$ *to the dominant eigenvalue* λ_1*.*

Proof Using (2.5) it follows that

$$\mathbf{y}^{(k)} = \mathbf{M}^k\,\mathbf{y}^{(0)} = \sum_{i=1}^{n} \lambda_i^k \mathbf{v}_i \underbrace{\mathbf{v}_i^T \mathbf{y}^{(0)}}_{=\alpha_i} = \sum_{i=1}^{n} \lambda_i^k \alpha_i \mathbf{v}_i \,.$$

Since $\mathbf{y}^{(0)}$ is not orthogonal to $\mathbf{v}_1$, $\alpha_1 \neq 0$ holds and thus

$$\begin{aligned}
\lim_{k\mapsto\infty} \mathbf{x}^{(k)} &= \lim_{k\mapsto\infty} \frac{\sum_{i=1}^n \lambda_i^k \alpha_i \mathbf{v}_i}{\|\sum_{i=1}^n \lambda_i^k \alpha_i \mathbf{v}_i\|} = \lim_{k\mapsto\infty} \frac{\lambda_1^k \alpha_1 \mathbf{v}_1 + \sum_{i=2}^n \lambda_i^k \alpha_i \mathbf{v}_i}{\|\lambda_1^k \alpha_1 \mathbf{v}_1 + \sum_{i=2}^n \lambda_i^k \alpha_i \mathbf{v}_i\|} \\
&= \lim_{k\mapsto\infty} \frac{\lambda_1^k \quad \alpha_1 \mathbf{v}_1 + \sum_{i=2}^n (\lambda_i/\lambda_1)^k \alpha_i \mathbf{v}_i}{|\lambda_1^k| \;\|\alpha_1 \mathbf{v}_1 + \sum_{i=2}^n (\lambda_i/\lambda_1)^k \alpha_i \mathbf{v}_i\|} = \lim_{k\mapsto\infty} \frac{\lambda_1^k \alpha_1 \mathbf{v}_1}{|\lambda_1^k||\alpha_1|\|\mathbf{v}_1\|} \,,
\end{aligned}$$

which shows that $\mathbf{x}^{(k)}$ becomes asymptotically parallel to $\mathbf{v}_1$. Hence, we obtain $\left|\mathbf{v}_1^T \lim_{k\mapsto\infty} \mathbf{x}^{(k)}\right| = 1$. Moreover,

$$\begin{aligned}
\lim_{k\mapsto\infty} \mu^{(k)} &= \lim_{k\mapsto\infty} \frac{(\mathbf{y}^{(0)})^T \mathbf{M}^{k-1} \mathbf{M}^k \mathbf{y}^{(0)}}{\|\mathbf{M}^{k-1}\mathbf{y}^{(0)}\|^2} = \lim_{k\mapsto\infty} \frac{\sum_{i=1}^n \lambda_i^{2k-1} \alpha_i^2}{\sum_{i=1}^n \lambda_i^{2k-2} \alpha_i^2} \\
&= \lambda_1 \lim_{k\mapsto\infty} \frac{\alpha_1 + \sum_{i=2}^n (\lambda_i/\lambda_1)^{2k-1} \alpha_i^2}{\alpha_1 + \sum_{i=2}^n (\lambda_i/\lambda_1)^{2k-2} \alpha_i^2} = \lambda_1 \,,
\end{aligned}$$

which completes the proof. ■

Once the dominant eigenvalue λ_1 and the corresponding eigenvector $\mathbf{v}_1$ are determined, we can compute

$$\mathbf{M}_1 = \mathbf{M} - \hat{\lambda}_1 \hat{\mathbf{v}}_1 \hat{\mathbf{v}}_1^T = \left(\lambda_1 \mathbf{v}_1 \mathbf{v}_1^T - \hat{\lambda}_1 \hat{\mathbf{v}}_1 \hat{\mathbf{v}}_1^T\right) + \sum_{i=2}^{n} \lambda_i \mathbf{v}_i \mathbf{v}_i^T$$

and apply the power iteration to $\mathbf{M}_1$ for computing λ_2, provided the assumptions of Theorem 2.9 are fulfilled for λ_2. This can be iterated further as long as $|\lambda_t| > |\lambda_{t-1}|$. Care must be taken about numerical errors which build up iteratively because of the deviation between the numerically approximated and the true values.

The computation of eigenvalues of big matrices with millions of entries can be very time consuming and inefficient. A useful method for at least bounding the eigenvalues by the sum of nondiagonal elements is provided by Gershgorin's theorem, cf. [16]. We present the theorem in its general form, also covering complex matrices and eigenvalues.

Theorem 2.10 (S. A. Gershgorin, 1931) *Let* $\mathbf{M} = \left(m_{ij}\right) \in \mathbb{C}^{n \times n}$ *with spectrum* $\mathcal{S} = \{\lambda \in \mathbb{C} \mid \det(\mathbf{M} - \lambda \mathbf{I}_n) = 0\}$. *For* $i = 1, \ldots, n$ *define the Gershgorin circles*

$$\mathcal{R}_i = \left\{ z \in \mathbb{C} \;\middle|\; |z - m_{ii}| \leq \sum_{\substack{j=1 \\ j \neq i}}^{n} |m_{ij}| \right\} \tag{2.8}$$

and

$$\mathcal{C}_j = \left\{ z \in \mathbb{C} \;\middle|\; |z - m_{jj}| \leq \sum_{\substack{i=1 \\ i \neq j}}^{n} |m_{ij}| \right\}.$$

Then it holds that

$$\mathcal{S} \subseteq \bigcup_{i=1}^{n} (\mathcal{R}_i \cap \mathcal{C}_i),$$

i.e., all eigenvalues of $\mathbf{M}$ *are contained in at least one of the intersection of row- and column-wise Gershgorin circles.*

Proof Let $\lambda \in \mathcal{S}$ be an eigenvalue of $\mathbf{M}$ corresponding to the eigenvector $\mathbf{v} = (v_1, v_2, \ldots, v_n)^T$. Let i be the index such that $|v_i| = \max_{1 \leq j \leq n} |v_j|$. Then from the eigenvalue equation $\lambda \mathbf{v} = \mathbf{M}\mathbf{v}$ we conclude that $(\lambda - m_{ii})v_i = \sum_{\substack{j=1 \\ j \neq i}}^{n} m_{ij} v_j$ such that because of the triangle inequality

$$|\lambda - m_{ii}|\, |v_i| = \left| \sum_{\substack{j=1 \\ j \neq i}}^{n} m_{ij} v_j \right| \leq \sum_{\substack{j=1 \\ j \neq i}}^{n} |m_{ij}|\, |v_j|.$$

Dividing both sides by $|v_i|$ yields

$$|\lambda - m_{ii}| \leq \sum_{\substack{j=1 \\ j \neq i}}^{n} |m_{ij}| \underbrace{|v_j/v_i|}_{\leq 1} \leq \sum_{\substack{j=1 \\ j \neq i}}^{n} |m_{ij}|,$$

which shows that $\lambda \in \mathcal{R}_i$. Analogously, applying the same steps to $\mathbf{M}^T$ gives $\lambda \in C_i$ and hence $\lambda \in \mathcal{R}_i \cap C_i$, which completes the proof. ■

Notation becomes easier if matrix $\mathbf{M} \in \mathbb{R}^{n \times n}$ is symmetric since the Gershgorin circles $\mathcal{R}_i$ and C_i coincide and all eigenvalues are real.

Corollary 2.11 *If* $\mathbf{M} = (m_{ij}) \in \mathbb{R}^{n \times n}$ *is symmetric, then every eigenvalue of* $\mathbf{M}$ *lies within at least one of the intervals,* $i = 1, \ldots, n$,

$$\Big[m_{ii} - \sum_{\substack{j=1 \\ j \neq i}} m_{ij},\ m_{ii} + \sum_{\substack{j=1 \\ j \neq i}} m_{ij}\Big].$$

The following example demonstrates how an upper and a lower bound for the eigenvalues can be derived by the Gerschgorin intervals.

Example 2.12 Matrix

$$\mathbf{M} = \begin{pmatrix} 2 & -1 & 0 \\ -1 & 3 & -1 \\ 0 & -1 & 4 \end{pmatrix}$$

has the spectrum $\mathcal{S} = \{1.27, 3.00, 4.73\}$. The Gerschgorin bounds are derived from $[1, 3] \cup [1, 5] \cup [3, 5] = [1, 5]$, such that $1 \leq \lambda_i \leq 5$ for all $\lambda_i \in \mathcal{S}$ can be derived without computing the full spectrum. The bounds also show that $\mathbf{M}$ is positive definite.

2.2 Matrix Norms, Trace and Partitioned Matrices

The set of $m \times n$ matrices is obviously a linear vector space. Addition of matrices and multiplication by scalars is clearly defined. In this section we will endow this linear space by more structure. The trace operator will serve as a tool to introduce an inner product and a norm. This will allow for the concept of *distance* between matrices. Matrix computations become often much easier if the concept of partitioned matrices and corresponding calculus are available, so that we also introduce this basic tool.

We commence by introducing two different norms, both of great importance for low-dimensional space approximations in data analytics.

Definition 2.13 (*Matrix norms*)

(a) For any $\mathbf{M} = (m_{ij}) \in \mathbb{R}^{m \times n}$

$$\|\mathbf{M}\|_{\mathrm{F}} = \left(\sum_{i=1}^{m} \sum_{j=1}^{n} m_{ij}^2 \right)^{1/2}$$

is called the *Frobenius norm* of $\mathbf{M}$.

(b) If $\mathbf{M} \in \mathbb{R}^{n \times n}$ is symmetric with eigenvalues $\lambda_1, \ldots, \lambda_n$, then

$$\|\mathbf{M}\|_{\mathrm{S}} = \max_{1 \le i \le n} |\lambda_i|$$

is called the *spectral norm* of $\mathbf{M}$.

Both the Frobenius and the spectral norm have the following characteristic properties. By $\| \cdot \|_{\mathrm{F/S}}$ we denote either of them.

Lemma 2.14 *The following holds for the Frobenius and the spectral norm:*

(a) For any real vector $\mathbf{x}$ *and any real matrix* $\mathbf{M}$ *of appropriate dimension*

$$\|\mathbf{M}\mathbf{x}\|_2 \le \|\mathbf{M}\|_{F/S} \, \|\mathbf{x}\|_2 \, ,$$

(b) For any two matrices $\mathbf{M}_1$ *and* $\mathbf{M}_2$ *of appropriate dimension*

$$\|\mathbf{M}_1\mathbf{M}_2\|_{F/S} \le \|\mathbf{M}_1\|_{F/S} \, \|\mathbf{M}_2\|_{F/S} \, ,$$

(c) For any real matrix $\mathbf{M}$ *and real orthogonal matrices* $\mathbf{U}$ *and* $\mathbf{W}$ *of appropriate dimension*

$$\|\mathbf{U}\mathbf{M}\mathbf{W}\|_{F/S} = \|\mathbf{M}\|_{F/S}.$$

The proof is left as an exercise, see Exercise 2.3.

We next introduce the trace of a matrix and some of its properties.

Definition 2.15 The *trace* of a matrix $\mathbf{M} = \left(m_{ij}\right) \in \mathbb{R}^{n \times n}$ is defined as

$$\operatorname{tr}(\mathbf{M}) = \sum_{i=1}^{n} m_{ii} \tag{2.9}$$

The following properties of the trace will be frequently used.

Lemma 2.16 *(a) The trace is commutative, i.e., for matrices* $\mathbf{A} \in \mathbb{R}^{m \times n}$ *and* $\mathbf{B} \in \mathbb{R}^{n \times m}$ *it holds that*

$$\operatorname{tr}(\mathbf{A}\mathbf{B}) = \operatorname{tr}(\mathbf{B}\mathbf{A}).$$

(b) The trace is linear, i.e., for matrices $\mathbf{A}, \mathbf{B} \in \mathbb{R}^{n \times n}$ *and scalars* $\alpha, \beta \in \mathbb{R}$ *it holds that*

$$\operatorname{tr}(\alpha\mathbf{A} + \beta\mathbf{B}) = \alpha \operatorname{tr}(\mathbf{A}) + \beta \operatorname{tr}(\mathbf{B}).$$

(c) *For any symmetric matrix* $\mathbf{M}$ *with eigenvalues* $\lambda_1, \lambda_2, \ldots, \lambda_n$ *the following holds:*

$$\operatorname{tr}(\mathbf{M}) = \sum_{i=1}^{n} \lambda_i \quad \textit{and} \quad \det(\mathbf{M}) = \prod_{i=1}^{n} \lambda_i \,.$$

(d) *For* $\mathbf{M} \in \mathbb{R}^{m \times n}$ *it holds that*

$$\operatorname{tr}\big(\mathbf{M}^T\mathbf{M}\big) = \|\mathbf{M}\|_F^2 \,.$$

The proof of the above elementary properties is again left to the reader, see Exercise 2.4.

We conclude this chapter by considering the inverse of a partitioned matrix. Extensions to general matrices and a proof of Theorem 2.17 may be found, e.g., in [30].

Theorem 2.17 *Consider the symmetric and invertible block matrix* $\mathbf{M} \in \mathbb{R}^{n \times n}$ *partitioned into invertible matrix* $\mathbf{A} \in \mathbb{R}^{m \times m}$, $\mathbf{B} \in \mathbb{R}^{m \times (n-m)}$ *and* $\mathbf{C} \in \mathbb{R}^{(n-m) \times (n-m)}$, $m \le n$, *as*

$$\mathbf{M} = \begin{pmatrix} \mathbf{A} & \mathbf{B} \\ \mathbf{B}^T & \mathbf{C} \end{pmatrix} .$$

Then

$$\mathbf{M}^{-1} = \begin{pmatrix} \mathbf{A}^{-1} + \mathbf{F}\mathbf{E}^{-1}\mathbf{F}^T & -\mathbf{F}\mathbf{E}^{-1} \\ -\mathbf{E}^{-1}\mathbf{F}^T & \mathbf{E}^{-1} \end{pmatrix}$$

and

$$\det(\mathbf{M}) = \det(\mathbf{A}) \det(\mathbf{E}) ,$$

where the matrix

$$\mathbf{E} = \mathbf{C} - \mathbf{B}^T\mathbf{A}^{-1}\mathbf{B}$$

is the called the Schur complement *of* $\mathbf{M}$.

2.3 Matrix Ordering and Matrix Monotone Functions

The concept of monotonicity for real functions with real arguments inevitably needs the ordering of the arguments. We call a real function $f : \mathbb{R} \to \mathbb{R}$ monotonically increasing if $f(a) \le f(b)$ for all $a \le b \in \mathbb{R}$.

In this section we will carry over monotonicity to real functions defined on the set of matrices. We cannot expect that a *full ordering* on the set of matrices exists, in the sense that any two of them are comparable. However, if there is an ordering which at least applies to selected matrices, then monotonicity can be partially defined. Such orderings are called *semi-orderings*, saying that not necessarily all elements of a set can be compared w.r.t. the underlying ordering.

The simple example of ordering matrices (or vectors) by comparing their real components "$\mathbf{A} \leq \mathbf{B}$ if $a_{ij} \leq b_{ij}$ for all entries" demonstrates that only selected matrices are comparable. More sophisticated semi-orderings are needed for data analytics. A most useful one was introduced by Karl Löwner [23] in the class of symmetric matrices.

Definition 2.18 (*Loewner semi-ordering, 1934*) Let $\mathbf{V}, \mathbf{W} \in \mathbb{R}^{n \times n}$ be symmetric matrices. The Loewner semi-ordering $\geq_{\mathrm{L}}$ is defined by

$$\mathbf{V} \geq_{\mathrm{L}} \mathbf{W} \text{ if } \mathbf{V} - \mathbf{W} \text{ is nonnegative definite.}$$

Correspondingly we write $\mathbf{V} \geq_{\mathrm{L}} \mathbf{0}$ if $\mathbf{V}$ is nonnegative definite. The reversed symbol $\mathbf{V} \leq_{\mathrm{L}} \mathbf{W}$ has the obvious meaning that $\mathbf{W} - \mathbf{V} \geq_{\mathrm{L}} \mathbf{0}$. The extension to positive definite matrices is immediate. We use $\mathbf{V} >_{\mathrm{L}} \mathbf{W}$ if $\mathbf{V} - \mathbf{W}$ is positive definite.

For any symmetric matrices $\mathbf{U}, \mathbf{V}, \mathbf{W} \in \mathbb{R}^{n \times n}$ and $\alpha \in \mathbb{R}_+$ the Loewner semi-ordering has the following properties:

(i) Reflexivity: $\mathbf{U} \leq_{\mathrm{L}} \mathbf{U}$
(ii) Antisymmetry: if $\mathbf{U} \leq_{\mathrm{L}} \mathbf{V}$ and $\mathbf{V} \leq_{\mathrm{L}} \mathbf{U}$ then $\mathbf{U} = \mathbf{V}$
(iii) Transitivity: if $\mathbf{U} \leq_{\mathrm{L}} \mathbf{V}$ and $\mathbf{V} \leq_{\mathrm{L}} \mathbf{W}$ then $\mathbf{U} \leq_{\mathrm{L}} \mathbf{W}$
(iv) Additivity: if $\mathbf{U} \leq_{\mathrm{L}} \mathbf{V}$ and $\mathbf{W} \geq_{\mathrm{L}} \mathbf{0}$ then $\mathbf{U} + \mathbf{W} \leq_{\mathrm{L}} \mathbf{V} + \mathbf{W}$
(v) Scalability: if $\mathbf{U} \leq_{\mathrm{L}} \mathbf{V}$ then $\alpha\mathbf{U} \leq_{\mathrm{L}} \alpha\mathbf{V}$

The trace and the determinant of matrices are obviously real valued function. By the Loewner semi-ordering the concept of monotonicity (and even convexity) becomes available, as we will see next.

Theorem 2.19 *Given* $\mathbf{V} = (v_{ij}) \geq_{\mathrm{L}} \mathbf{0}$ *and* $\mathbf{W} = (w_{ij}) \geq_{\mathrm{L}} \mathbf{0}$ *with* $\mathbf{V} \leq_{\mathrm{L}} \mathbf{W}$ *and eigenvalues* $\lambda_1(\mathbf{V}) \geq \lambda_2(\mathbf{V}) \geq \cdots \geq \lambda_n(\mathbf{V})$ *and* $\lambda_1(\mathbf{W}) \geq \lambda_2(\mathbf{W}) \geq \cdots \geq \lambda_n(\mathbf{W})$. *Then the following inequalities hold:*

(a) $\lambda_i(\mathbf{V}) \leq \lambda_i(\mathbf{W})$ *for all* $i = 1, \ldots, n$
(b) $v_{ii} \leq w_{i,i}$ *for all* $i = 1, \ldots, n$
(c) $v_{ii} + v_{jj} - 2v_{ij} \leq w_{ii} + w_{jj} - 2w_{ij}$ *for all* $i, j = 1, \ldots, n$
(d) $\mathrm{tr}(\mathbf{V}) \leq \mathrm{tr}(\mathbf{W})$
(e) $\det(\mathbf{V}) \leq \det(\mathbf{W})$
(f) $\mathrm{tr}(\mathbf{MV}) \leq \mathrm{tr}(\mathbf{MW})$ *for any* $\mathbf{M} \geq_{\mathrm{L}} \mathbf{0}$

Once (a) is proved, the other assertions follow easily. For a complete proof we refer to the book [27].

Example 2.20 For a matrix $\mathbf{M} = \begin{pmatrix} \alpha & \beta \\ \beta & \gamma \end{pmatrix}$ with entries $\alpha, \beta, \gamma \in \mathbb{R}$, the eigenvalues are given by

$$\lambda_{1/2}(\mathbf{M}) = \frac{\alpha + \gamma}{2} \pm \frac{1}{2}\sqrt{(\alpha - \gamma)^2 + 4\beta^2}$$

(see Exercise 2.7). We consider the matrices

$$\mathbf{A} = \begin{pmatrix} 4 & 2 \\ 2 & 1 \end{pmatrix}, \quad \mathbf{B} = \begin{pmatrix} 5 & 2 \\ 2 & 2 \end{pmatrix}, \quad \mathbf{C} = \begin{pmatrix} 2 & 2 \\ 2 & 5 \end{pmatrix}.$$

By the above formula the eigenvalues are $\lambda_1(\mathbf{A}) = 5$, $\lambda_2(\mathbf{A}) = 0$, $\lambda_1(\mathbf{B}) = \lambda_1(\mathbf{C}) = 6$, and $\lambda_2(\mathbf{B}) = \lambda_2(\mathbf{C}) = 1$. Hence, $\mathbf{A} \geq_{\mathrm{L}} \mathbf{0}$, $\mathbf{B} >_{\mathrm{L}} \mathbf{0}$, and $\mathbf{C} >_{\mathrm{L}} \mathbf{0}$.

The differences are

$$\mathbf{B} - \mathbf{A} = \begin{pmatrix} 1 & 0 \\ 0 & 1 \end{pmatrix}, \quad \mathbf{C} - \mathbf{A} = \begin{pmatrix} -2 & 0 \\ 0 & 4 \end{pmatrix}, \quad \text{and} \quad \mathbf{C} - \mathbf{B} = \begin{pmatrix} -3 & 0 \\ 0 & 3 \end{pmatrix},$$

such that $\mathbf{B} - \mathbf{A}$ is nonnegative definite, hence, $\mathbf{B} \geq_{\mathrm{L}} \mathbf{A}$. Both matrices $\mathbf{C} - \mathbf{A}$ and $\mathbf{C} - \mathbf{B}$ are indefinite such that matrix $\mathbf{C}$ is not comparable with neither $\mathbf{A}$ nor $\mathbf{B}$. This demonstrates that $\geq_{\mathrm{L}}$ is a semi-ordering. Moreover, since $\mathbf{B} \geq_{\mathrm{L}} \mathbf{A}$, all inequalities in Theorem 2.19 are satisfied by $\mathbf{A}$ and $\mathbf{B}$. According to the last inequality in Theorem 2.19 we also get

$$\operatorname{tr}(\mathbf{CA}) = \operatorname{tr} \begin{pmatrix} 12 & 6 \\ 18 & 9 \end{pmatrix} = 21 \text{ and } \operatorname{tr}(\mathbf{CB}) = \operatorname{tr} \begin{pmatrix} 14 & 8 \\ 20 & 14 \end{pmatrix} = 28.$$

2.4 Stochastic Matrices and Permutation Matrices

Random walks on graphs are a useful tool in data analytics, as we will demonstrate for *diffusion maps* in Chap. 4. Random transitions between states are described by stochastic matrices, which are square matrices with nonnegative entries and row sums equal to one. Each row corresponds to a discrete probability distribution, describing how the random walker selects the next state to visit. A special case of stochastic matrices are permutation matrices, which will also be introduced in this chapter. Stochastic matrices play an important role in matrix optimization problems.

The intention of this section is to carefully prepare understanding of matrix optimization for the purpose of dimensionality reduction in Chap. 4 and Markovian structures for machine learning in Chap. 7. We want to build a solid foundation for data analytical problems students may encounter in the future.

Definition 2.21 A rectangular matrix $\mathbf{M} = (m_{ij})_{\substack{1 \leq i \leq m \\ 1 \leq j \leq n}}$ is called a *stochastic matrix* or a *transition matrix* if $m_{ij} \geq 0$ for all i, j and $\sum_{j=1}^{n} m_{ij} = 1$ for all i.

For stochastic matrices $\mathbf{M} \in \mathbb{R}^{m \times n}$ it holds that $\mathbf{M1}_n = \mathbf{1}_m$. Therefor, multiplication from the left $\mathbf{pM} = \mathbf{q}$ with a stochastic row vector $\mathbf{p} \in \mathbb{R}^m$ yields a stochastic row vector $\mathbf{q} \in \mathbb{R}^n$.

Furthermore, the set of stochastic matrices is convex since $\alpha\mathbf{M}_1 + (1 - \alpha)\mathbf{M}_2$ is a stochastic matrix for any stochastic matrices $\mathbf{M}_1$ and $\mathbf{M}_2$ and $\alpha \in [0, 1]$.

Lemma 2.22 *Let* $\mathbf{M} \in \mathbb{R}^{n \times n}$ *be a stochastic matrix. The all-one vector* $\mathbf{1}_n$ *is an eigenvector of* $\mathbf{M}$ *corresponding to the eigenvalue* 1. *It holds that* $|\lambda| \leq 1$ *for all eigenvalues eigenvalues* λ *of* $\mathbf{M}$.

Proof By definition, $\mathbf{M}\mathbf{1}_n = \mathbf{1}_n$ holds such that 1 is an eigenvalue with eigenvector $\mathbf{1}_n$. Since $\mathbf{M}$ is real, the Gershgorin circles in (2.8) are

$$\begin{aligned} \mathcal{R}_i &= \Big\{ z \in \mathbb{C} \;\Big|\; |z - m_{ii}| \leq \sum_{\substack{j=1\\ j\neq i}}^{n} |m_{i,j}| \Big\} \\ &\subseteq \Big\{ z \in \mathbb{C} \;\Big|\; |z| \leq m_{ii} + \sum_{\substack{j=1\\ j\neq i}}^{n} m_{ij} \Big\} = \big\{ z \in \mathbb{C} \;\big|\; |z| \leq 1 \big\},\; i = 1, \ldots, n, \end{aligned}$$

which shows that the spectrum of $\mathbf{M}$ lies in the unit disc of the complex plane. ■

Definition 2.23 A square matrix $\mathbf{M} = (m_{ij})_{\substack{1\leq i\leq n\\ 1\leq j\leq n}}$ is called *doubly stochastic* if $\mathbf{M}$ and $\mathbf{M}^T$ are stochastic matrices, i.e., $m_{ij} \geq 0$ for all i, j, $\sum_{j=1}^{n} m_{ij} = 1$ for all i and $\sum_{i=1}^{n} m_{ij} = 1$ for all j.

A square stochastic matrix is called a *permutation matrix* if each row and each column consists of a single one and zeros otherwise.

For a doubly stochastic matrix $\mathbf{M}$ the relations $\mathbf{1}^T\mathbf{M} = \mathbf{1}^T$ and $\mathbf{M}\mathbf{1} = \mathbf{1}$ hold such that 1 is a right and left eigenvalue with all-one eigenvectors.

Obviously permutation matrices are doubly stochastic.

Example 2.24 There are $n!$ permutation matrices of size $n \times n$. For example, there are $3! = 6$ permutation matrices of size 3×3, namely

$$\begin{pmatrix} 1&0&0\\0&1&0\\0&0&1 \end{pmatrix}, \begin{pmatrix} 1&0&0\\0&0&1\\0&1&0 \end{pmatrix}, \begin{pmatrix} 0&1&0\\1&0&0\\0&0&1 \end{pmatrix}, \begin{pmatrix} 0&1&0\\0&0&1\\1&0&0 \end{pmatrix}, \begin{pmatrix} 0&0&1\\1&0&0\\0&1&0 \end{pmatrix}, \begin{pmatrix} 0&0&1\\0&1&0\\1&0&0 \end{pmatrix},$$

as can be easily checked.

Permutation matrices are named as such since right multiplication of a matrix by a permutation matrix interchanges the order of columns. Accordingly, left multiplication interchanges the rows.

The set of doubly stochastic matrices is convex since $\alpha\mathbf{M}_1 + (1-\alpha)\mathbf{M}_2$ is doubly stochastic for any doubly stochastic matrices $\mathbf{M}_1, \mathbf{M}_2$ and $\alpha \in [0, 1]$.

Any convex combination of permutation matrices yields a doubly stochastic matrix, as can be easily seen. The reversed question, whether each stochastic matrix can be written as a convex combination of permutation matrices, is answered by the following famous theorem of G. Birkhoff [6].

Theorem 2.25 (Extreme points of doubly stochastic matrices) *The permutation matrices constitute the extreme points of the set of stochastic matrices. This means that each stochastic matrix* $\mathbf{M} \in \mathbb{R}^{n \times n}$ *has a representation as*

$$\mathbf{M} = \sum_{i=1}^{r} \alpha_i \mathbf{P}_i$$

with permutation matrices $\mathbf{P}_i$ *and appropriate* $\alpha_i \geq 0$ *with* $\sum_{i=1}^{r} \alpha_i = 1$.

Proof Obviously, a convex combination of permutation matrices is a doubly stochastic matrix.

Now suppose that $\mathbf{M} = \big(m_{k\ell}\big)_{k,\ell} \in \mathbb{R}^{n \times n}$ is a doubly stochastic matrix, but not a permutation matrix. Then there are at least four entries $m_{k_1\ell_1}, m_{k_1\ell_2}, m_{k_2\ell_1}, m_{k_2\ell_2}$, which are neither one nor zero, $1 \leq k_1, k_2, \ell_1, \ell_2 \leq n$. Without loss of generality we assume that

$$u = m_{k_1\ell_1} = \min\big\{m_{k_1\ell_1}, m_{k_1\ell_2}, m_{k_2\ell_1}, m_{k_2\ell_2}\big\}. \tag{2.10}$$

Otherwise, indices are rearranged such that (2.10) holds. We then define

$$v = \min\big\{m_{k_1\ell_2}, m_{k_2\ell_1}\big\}.$$

From $\mathbf{M}$ two doubly stochastic matrices $\mathbf{M}' = \big(m'_{k\ell}\big)_{k,\ell}$ and $\mathbf{M}'' = \big(m''_{k\ell}\big)_{k,\ell}$ are constructed by replacing the four entries by

$$m'_{k_1\ell_1} = m_{k_1\ell_1} - u, \qquad m'_{k_1\ell_2} = m_{k_1\ell_2} + u,$$
$$m'_{k_2\ell_1} = m_{k_2\ell_1} + u, \qquad m'_{k_2\ell_2} = m_{k_2\ell_2} - u$$

and

$$m''_{k_1\ell_1} = m_{k_1\ell_1} + v, \qquad m''_{k_1\ell_2} = m_{k_1\ell_2} - v,$$
$$m''_{k_2\ell_1} = m_{k_2\ell_1} - v, \qquad m''_{k_2\ell_2} = m_{k_2\ell_2} + v,$$

and unaltered entries $m'_{k\ell} = m''_{k\ell} = m_{k\ell}$ otherwise. Defining $\alpha = \frac{v}{u+v}$ with $0 < \alpha < 1$ yields the identity

$$\mathbf{M} = \alpha\mathbf{M}' + (1 - \alpha)\mathbf{M}'' \tag{2.11}$$

such that $\mathbf{M}$ is written as a proper convex combination of two doubly stochastic matrices $\mathbf{M}_1 \neq \mathbf{M}_2$. This shows that $\mathbf{M}$ cannot be an extreme point of the set of doubly stochastic matrices.

Moreover, both new matrices $\mathbf{M}_1$ and $\mathbf{M}_2$ have at least one more zero entry than $\mathbf{M}$. Recursively applying the same procedure to both $\mathbf{M}'$ and $\mathbf{M}''$ until a permutation matrix is achieved yields finally a set of permutation matrices and convex weights $\alpha_i \geq 0$ such that $\mathbf{M}$ can be written as a convex combination of permutation matrices.

The weights are nonnegative and by construction add to one. The recursion above stops latest after $r = (n-1)^2 + 1$ steps. ■

Since any column (row) of a doubly stochastic matrix is determined by the difference between the all-one vector and the sum of the remaining columns (rows), the number $(n-1)^2 + 1$ of convex combinations results from the number $(n-1)^2$ of remaining matrix elements plus an additional addend for the discarded matrix elements. A detailed discussion of the best lower bound on the number of necessary extreme points is given in [14].

2.5 Extremals of Matrix Functions and Matrix Approximation

Real valued functions of matrices will be investigated in this section. Finding maxima and minima of such functions will allow us to determine low dimensional approximations to data in Chap. 4. In many cases, extrema of real valued matrix functions can be determined by their eigenvalues and spaces spanned by the corresponding eigenvectors.

We commence by introducing the *Courant-Fischer min-max principle* which characterizes the eigenvalues and singular values of matrices by iterated extrema of quadratic forms. It was first discussed by Ernst Fischer [15], and used later by Richard Courant [11] for describing eigenvalues of differential equations.

Theorem 2.26 (E. Fischer, 1905, and R. Courant, 1924) *If* $\mathbf{M} \in \mathbb{R}^{n\times n}$ *is a symmetric matrix with eigenvalues* $\lambda_1 \geq \lambda_2 \geq \cdots \geq \lambda_n$*, then for all* $k = 1, \ldots, n$

$$\max_{\substack{\mathcal{V}\subseteq\mathbb{R}^n\\ \dim(\mathcal{V})=k}} \min_{\substack{\mathbf{x}\in\mathcal{V},\\ \|\mathbf{x}\|_2=1}} \mathbf{x}^T\mathbf{M}\mathbf{x} = \lambda_k \tag{2.12}$$

and

$$\min_{\substack{\mathcal{V}\subseteq\mathbb{R}^n\\ \dim(\mathcal{V})=n-k+1}} \max_{\substack{\mathbf{x}\in\mathcal{V},\\ \|\mathbf{x}\|_2=1}} \mathbf{x}^T\mathbf{M}\mathbf{x} = \lambda_k \tag{2.13}$$

where $\mathcal{V}$ *denotes linear subspaces. In both equations, the optimum is attained at* $\mathbf{x}^* = \mathbf{v}_k$*, where* $\mathbf{v}_k$ *is the normalized eigenvector corresponding to eigenvalue* λ_k*.*

Proof We only prove (2.12), the proof of (2.13) is analogous.

Let $M = \mathbf{V}\boldsymbol{\Lambda}\mathbf{V}^T$ be the spectral decomposition of $\mathbf{M} \in \mathbb{R}^{n\times n}$ with eigenvalues $\lambda_1 \geq \ldots \lambda_n$ and corresponding eigenvectors forming the orthogonal matrix $\mathbf{V} = (\mathbf{v}_1, \ldots, \mathbf{v}_n)$.

For $1 \leq k \leq n$ we define

$$\mathcal{S} = \text{Span}\{\mathbf{v}_1, \ldots, \mathbf{v}_k\} \text{ and } \mathcal{T} = \text{Span}\{\mathbf{v}_k, \ldots, \mathbf{v}_n\}.$$

For all $\mathbf{w} = \sum_{i=1}^{k} c_i \mathbf{v}_i \in \mathcal{S}$ it holds that

$$\begin{aligned}\mathbf{w}^T\mathbf{M}\mathbf{w} &= \Big(\sum_{i=1}^{k} c_i\mathbf{v}_i\Big)^T \Big(\sum_{i=1}^{k} c_i\mathbf{M}\mathbf{v}_i\Big)\\ &= \Big(\sum_{i=1}^{k} c_i\mathbf{v}_i\Big)^T \Big(\sum_{i=1}^{k} c_i\lambda_i\mathbf{v}_i\Big)\\ &= \sum_{i=1}^{n} c_i^2\lambda_i \geq \lambda_k \sum_{i=1}^{k} c_i^2 = \lambda_k \mathbf{w}^T\mathbf{w}\,.\end{aligned} \tag{2.14}$$

Analogously we get for any $\mathbf{w} = \sum_{i=1}^{k} c_i \mathbf{v}_i \in \mathcal{T}$ that

$$\mathbf{w}^T\mathbf{M}\mathbf{w} \leq \lambda_k \mathbf{w}^T\mathbf{w}\,. \tag{2.15}$$

We begin the proof by first noting that Eq. (2.12) is equivalent to

$$\max_{\substack{\mathcal{V}\subseteq\mathbb{R}^n\\ \dim(\mathcal{V})=k}} \min_{\substack{\mathbf{x}\in\mathcal{V}\\ \mathbf{x}\neq\mathbf{0}}} \frac{\mathbf{x}^T\mathbf{M}\mathbf{x}}{\mathbf{x}^T\mathbf{x}} = \mu_k\,. \tag{2.16}$$

By (2.14) a lower bound for (2.16) is given by

$$\mu_k \geq \max_{\mathcal{S}} \min_{\substack{\mathbf{x}\in\mathcal{S}\\ \mathbf{x}\neq\mathbf{0}}} \frac{\mathbf{x}^T\mathbf{M}\mathbf{x}}{\mathbf{x}^T\mathbf{x}} = \min_{\substack{\mathbf{x}\in\mathcal{S}\\ \mathbf{x}\neq\mathbf{0}}} \frac{\mathbf{x}^T\mathbf{M}\mathbf{x}}{\mathbf{x}^T\mathbf{x}} \geq \lambda_k$$

Because of the dimension formula

$$\dim(\mathcal{V}\cap\mathcal{T}) = \dim(\mathcal{V}) + \dim(\mathcal{T}) - \dim(\mathcal{V}+\mathcal{T}) \geq k + n - k + 1 - n = 1$$

there is at least one nonzero $\mathbf{w} \in \mathcal{V}\cap\mathcal{T}$. From (2.15) an upper bound for (2.16) is obtained as

$$\mu_k \leq \max_{\substack{\mathcal{V}\subseteq\mathbb{R}^n\\ \dim(\mathcal{V})=k}} \min_{\substack{\mathbf{x}\in\mathcal{V}\cap\mathcal{T}\\ \mathbf{x}\neq\mathbf{0}}} \frac{\mathbf{x}^T\mathbf{M}\mathbf{x}}{\mathbf{x}^T\mathbf{x}} \leq \lambda_k\,.$$

In summary, equality $\mu_k = \lambda_k$ follows, which shows (2.12).

The optima are attained at $\mathbf{x}^* = \mathbf{v}_k$ since $\mathbf{v}_k^T\mathbf{M}\mathbf{v}_k = \lambda_k$ holds by the definition of eigenvalues and eigenvectors. ■

The cases $k = 1$ and $k = n$ in Theorem 2.26 show that

$$\min_{\mathbf{x}\in\mathbb{R}^n,\,\mathbf{x}\neq\mathbf{0}} \frac{\mathbf{x}^T\mathbf{M}\mathbf{x}}{\mathbf{x}^T\mathbf{x}} = \lambda_n \quad\text{and}\quad \max_{\mathbf{x}\in\mathbb{R}^n,\,\mathbf{x}\neq\mathbf{0}} \frac{\mathbf{x}^T\mathbf{M}\mathbf{x}}{\mathbf{x}^T\mathbf{x}} = \lambda_1 \tag{2.17}$$

for any symmetric matrix $\mathbf{M} \in \mathbb{R}^{n\times n}$ with eigenvalues $\lambda_1 \geq \lambda_2 \geq \cdots \geq \lambda_n$. The ratio $\frac{\mathbf{x}^T\mathbf{M}\mathbf{x}}{\mathbf{x}^T\mathbf{x}}$ is called *Rayleigh-Ritz quotient*. It attains its minimum or maximum if $\mathbf{x}$ is the eigenvalue corresponding to the smallest or largest eigenvalue, respectively.

If $\mathbf{M}$ is nonnegative definite with nonnegative eigenvalues λ_i, the Rayleigh quotient has the following geometric interpretation. The hypersphere of vectors $\mathbf{x}$ with $\|\mathbf{x}\|_2 = 1$ is deformed into a hyperellipsoid by factor λ_i along the corresponding eigenvectors.

We next consider real valued functions of matrices and are particularly interested in finding extrema, cf. [13]. Typical examples are the trace and the closely connected Frobenius norm.

Theorem 2.27 (Ky Fan, 1949) *If* $\mathbf{M} \in \mathbb{R}^{n\times n}$ *is a symmetric matrix with eigenvalues* $\lambda_1 \geq \lambda_2 \geq \cdots \geq \lambda_n$ *and eigenvectors* $\mathbf{v}_1, \mathbf{v}_2, \ldots, \mathbf{v}_n$*, then for any* $1 \leq k \leq n$ *it holds that*

$$\max_{\substack{\mathbf{X}\subseteq\mathbb{R}^{n\times k}\\ \mathbf{X}^T\mathbf{X}=\mathbf{I}_k}} \operatorname{tr}\big(\mathbf{X}^T\mathbf{M}\mathbf{X}\big) = \sum_{i=1}^{k} \lambda_i \tag{2.18}$$

and

$$\min_{\substack{\mathbf{X}\subseteq\mathbb{R}^{n\times k}\\ \mathbf{X}^T\mathbf{X}=\mathbf{I}_k}} \operatorname{tr}\big(\mathbf{X}^T\mathbf{M}\mathbf{X}\big) = \sum_{i=1}^{k} \lambda_{n-i+1}\,. \tag{2.19}$$

The maximum in (2.18) is attained at $\mathbf{X}^* = (\mathbf{v}_1, \ldots, \mathbf{v}_k)$*, and the minimum in (2.19) is attained at* $\mathbf{X}^* = (\mathbf{v}_{n-k+1}, \ldots, \mathbf{v}_n)$*.*

Proof We only prove (2.18), the proof of (2.19) is analogous.

Let $\mathbf{X} = (\mathbf{x}_1, \mathbf{x}_2, \ldots, \mathbf{x}_k)$ with $\mathbf{X}^T\mathbf{X} = \mathbf{I}_k$ and $\mathbf{M} = \mathbf{V}\boldsymbol{\Lambda}\mathbf{V}^T$ the spectral decomposition with decreasing eigenvalues $\lambda_1, \ldots, \lambda_n$. Since $\mathbf{V}$ is orthogonal, we obtain

$$\operatorname{tr}\big(\mathbf{X}^T\mathbf{M}\mathbf{X}\big) = \operatorname{tr}\big(\mathbf{X}^T\mathbf{V}\boldsymbol{\Lambda}\mathbf{V}^T\mathbf{X}\big) = \operatorname{tr}\big(\mathbf{W}^T\boldsymbol{\Lambda}\mathbf{W}\big) = \sum_{i=1}^{k} \mathbf{w}_i^T\boldsymbol{\Lambda}\mathbf{w}_i\,,$$

with $\mathbf{W} = (\mathbf{w}_1, \ldots, \mathbf{w}_k) = \mathbf{V}^T\mathbf{X}$ and $\mathbf{W}^T\mathbf{W} = \mathbf{X}^T\mathbf{V}\mathbf{V}^T\mathbf{X} = \mathbf{I}_k$. In particular, it holds that $\|\mathbf{w}_i\|^2 = \sum_{j=1}^n w_{ji}^2 = 1$.

We next determine an upper bound for $\mathbf{w}_i^T\boldsymbol{\Lambda}\mathbf{w}_i$, $i = 1, \ldots, k$.

$$\begin{aligned}
\mathbf{w}_i^T\boldsymbol{\Lambda}\mathbf{w}_i &= \lambda_k \sum_{j=1}^{n} w_{ji}^2 + \sum_{j=1}^{k}(\lambda_j - \lambda_k)w_{ji}^2 + \sum_{j=k+1}^{n}(\lambda_j - \lambda_k)w_{ji}^2 \\
&\leq \lambda_k + \sum_{j=1}^{k}(\lambda_j - \lambda_k)w_{ji}^2\,.
\end{aligned}$$

Summing up over all $i = 1, \ldots, k$ and reverting the sign gives

$$-\sum_{i=1}^{k} \mathbf{w}_i^T \mathbf{\Lambda} \mathbf{w}_i \geq -\sum_{j=1}^{k} \lambda_k - \sum_{j=1}^{k}\sum_{i=1}^{k} (\lambda_j - \lambda_k) w_{ji}^2 \,.$$

Adding the sum of the k dominant eigenvalues yields

$$\sum_{i=1}^{k} \lambda_i - \sum_{i=1}^{k} \mathbf{w}_i^T \mathbf{\Lambda} \mathbf{w}_i \geq \sum_{j=1}^{k} (\lambda_j - \lambda_k) \left(1 - \sum_{i=1}^{k} w_{ji}^2\right) \geq 0$$

since the eigenvalues are ordered decreasingly and $\sum_{i=1}^{k} w_{ji}^2 = 1$ with nonnegative addends w_{ij}^2.

It is easy to see that equality holds for $\mathbf{X}^* = (\mathbf{v}_1, \ldots, \mathbf{v}_k)$, which completes the proof. ■

We will continue to derive lower and upper bounds on the trace of a product of two matrices. The following preparatory Lemma forms the key part of the proof. Extensions to multiple factors are given in [17].

Lemma 2.28 (G. Hardy, J. Littlewood, and G. Pólya, 1934) *For real numbers $x_1, \ldots, x_n$ and $y_1, \ldots, y_n$ the values in decreasing order are denoted by $x_{[1]} \geq \cdots \geq x_{[n]}$ and $y_{[1]} \geq \cdots \geq y_{[n]}$. Then*

$$\sum_{i=1}^{n} x_{[i]} y_{[n-i+1]} \leq \sum_{i=1}^{n} x_i y_i \leq \sum_{i=1}^{n} x_{[i]} y_{[i]} \,. \tag{2.20}$$

Proof Without loss of generality we may assume that $x_1 \geq \cdots \geq x_n$. If there is some index ℓ with $y_\ell < y_{\ell+1}$, then

$$\sum_{i=1}^{n} x_i y_i - \sum_{\substack{i=1\\ i\neq \ell,\ell+1}}^{n} x_i y_i - x_\ell y_{\ell+1} - x_{\ell+1} y_\ell = (x_\ell - x_{\ell+1})(y_\ell - y_{\ell+1}) \leq 0 \,.$$

Hence, by exchanging the values y_ℓ and $y_{\ell+1}$ the sum $\sum_{i=1}^{n} x_i y_i$ does not decrease. It even increases if $x_\ell > x_{\ell+1}$. By finitely many such exchange steps the right hand side of (2.20) is achieved as the maximum. Analogous arguments show that the left hand side is a lower bound. ■

Lemma 2.28 was used by Hans Richter [35] to determine a lower and an upper bound for the trace of a matrix product in terms of respective eigenvalues.

Theorem 2.29 (H. Richter, 1958) *If $\mathbf{A} \in \mathbb{R}^{n\times n}$ and $\mathbf{B} \in \mathbb{R}^{n\times n}$ are symmetric matrices with eigenvalues $\lambda_1 \geq \lambda_2 \geq \cdots \geq \lambda_n$ and $\mu_1 \geq \mu_2 \geq \cdots \geq \mu_n$, respectively, then*

$$\sum_{i=1}^{n} \lambda_i \mu_{n-i+1} \leq \operatorname{tr}(\mathbf{AB}) \leq \sum_{i=1}^{n} \lambda_i \mu_i \,. \tag{2.21}$$

Proof Both matrices $\mathbf{A}$ and $\mathbf{B}$ have a spectral decomposition $\mathbf{A} = \sum_{i=1}^{n} \lambda_i \mathbf{u}_i \mathbf{u}_i^T$ and $\mathbf{B} = \sum_{i=1}^{n} \mu_i \mathbf{v}_i \mathbf{v}_i^T$ with orthonormal eigenvectors $\mathbf{u}_i$ and $\mathbf{v}_i$, respectively. Hereby,

$$\mathrm{tr}(\mathbf{AB}) = \mathrm{tr}\Big(\sum_{i=1}^{n}\sum_{j=1}^{n} \lambda_i \mu_j \mathbf{u}_i \mathbf{u}_i^T \mathbf{v}_j \mathbf{v}_j^T\Big) = \sum_{i=1}^{n}\sum_{j=1}^{n} \lambda_i \mu_j \big(\mathbf{u}_i^T \mathbf{v}_j\big)^2 = \boldsymbol{\lambda}^T \mathbf{M} \boldsymbol{\mu},$$

where $\boldsymbol{\lambda} = (\lambda_1, \ldots, \lambda_n)^T$, $\boldsymbol{\mu} = (\mu_1, \ldots, \mu_n)^T$ and $\mathbf{M} = \big((\mathbf{u}_i^T \mathbf{v}_j)^2\big)_{1 \le i,j \le n}$. Note that matrix $\mathbf{M}$ is doubly stochastic, since $\big(\mathbf{u}_i^T \mathbf{v}_j\big)^2 \ge 0$ for all i, j, $\sum_{i=1}^{n} \big(\mathbf{u}_i^T \mathbf{v}_j\big)^2 = \mathbf{v}_j^T \big(\sum_{i=1}^{n} \mathbf{u}_i \mathbf{u}_i^T\big) \mathbf{v}_j = 1$ for all j, and similarly $\sum_{j=1}^{n} \big(\mathbf{u}_i^T \mathbf{v}_j\big)^2 = \mathbf{u}_i^T \big(\sum_{j=1}^{n} \mathbf{v}_j \mathbf{v}_j^T\big) \mathbf{u}_i = 1$ for all i, cf. Exercise 2.9. Hence, by Birkhoff's Theorem 2.25 $\mathbf{M}$ can be represented as a convex combination

$$\mathbf{M} = \sum_{k=1}^{m} \alpha_k \mathbf{P}_k \tag{2.22}$$

with $m \le (n-1)^2 + 1$, nonnegative $\alpha_1, \ldots, \alpha_m$ such that $\sum_{k=1}^{m} \alpha_k = 1$ and $\mathbf{P}_1, \ldots, \mathbf{P}_m \in \mathbb{R}^{n \times n}$ pairwise different permutation matrices. Using (2.22) we obtain

$$\mathrm{tr}(\mathbf{AB}) = \boldsymbol{\lambda}^T \Big(\sum_{k=1}^{m} \alpha_k \mathbf{P}_k\Big) \boldsymbol{\mu} = \sum_{k=1}^{m} \alpha_k \boldsymbol{\lambda}^T \mathbf{P}_k \boldsymbol{\mu}.$$

According to Lemma 2.28, each addend $\boldsymbol{\lambda}^T \mathbf{P}_k \boldsymbol{\mu}$ can be upper and lower bounded by

$$\sum_{i=1}^{n} \lambda_{[i]} \mu_{[n-i+1]} \le \boldsymbol{\lambda}^T \mathbf{P}_k \boldsymbol{\mu} \le \sum_{i=1}^{n} \lambda_{[i]} \mu_{[i]},$$

which completes the proof. ■

Matrix approximation problems play a prominent role in data analytics, particularly for dimensionality reduction. The following theorem is fundamental for Multidimensional Scaling, a method which seeks an embedding of points in a Euclidean space such that the pairwise distances between the points match given dissimilarities between objects, see Sect. 4.2. The following optimization problem was initially investigated by Carl Eckart and Gale Young [12].

Theorem 2.30 (C. Eckart and G. Young, 1936) *Let* $\mathbf{M} \in \mathbb{R}^{n \times n}$ *be a symmetric matrix with spectral decomposition* $\mathbf{V}\,\mathrm{diag}(\lambda_1, \lambda_2, \ldots, \lambda_n)\mathbf{V}^T$, $\lambda_1 \ge \lambda_2 \ge \cdots \ge \lambda_n$ *and* $1 \le k \le n$. *Then*

$$\min_{\mathbf{A} \succeq_L \mathbf{0},\, \mathrm{rank}(\mathbf{A}) \le k} \|\mathbf{M} - \mathbf{A}\|_F = \sum_{i=1}^{k} \big(\lambda_i - \lambda_i^+\big)^2 + \sum_{i=k+1}^{n} \lambda_i^2 \tag{2.23}$$

is attained at

$$\mathbf{A}^* = \mathbf{V}\,\mathrm{diag}(\lambda_1^+, \dots, \lambda_k^+, 0, \dots, 0)\mathbf{V}^T \tag{2.24}$$

where $\lambda_i^+ = \max\{\lambda_i, 0\}$ *denotes the positive part.*

Proof By Lemma 2.16 (d) the Frobenius norm can be written as

$$\|\mathbf{M}-\mathbf{A}\|_{\mathrm{F}} = \mathrm{tr}\big((\mathbf{M}-\mathbf{A})^T(\mathbf{M}-\mathbf{A})\big) = \mathrm{tr}(\mathbf{MM}) - 2\,\mathrm{tr}(\mathbf{MA}) + \mathrm{tr}(\mathbf{AA})\,. \tag{2.25}$$

Next, the trace of $\mathbf{M}^2$ is represented in terms of eigenvalues.

$$\mathrm{tr}\big(\mathbf{M}^2\big) = \mathrm{tr}\big(\mathbf{V}\boldsymbol{\Lambda}\mathbf{V}^T\mathbf{V}\boldsymbol{\Lambda}\mathbf{V}^T\big) = \mathrm{tr}\big(\mathbf{V}^T\mathbf{V}\boldsymbol{\Lambda}\mathbf{V}^T\mathbf{V}\boldsymbol{\Lambda}\big) = \mathrm{tr}\big(\boldsymbol{\Lambda}^2\big) = \sum_{i=1}^n \lambda_i^2$$

Analogously, it holds for the eigenvalues $\mu_1 \geq \dots \geq \mu_n$ of the symmetric matrix $\mathbf{A}$ that

$$\mathrm{tr}\big(\mathbf{A}^2\big) = \sum_{i=1}^n \mu_i^2\,.$$

A lower bound for (2.25) is obtained by Theorem 2.29.

$$\begin{aligned}\|\mathbf{M}-\mathbf{A}\|_{\mathrm{F}} &= \sum_{i=1}^n \lambda_i^2 - 2\,\mathrm{tr}(\mathbf{MA}) + \sum_{i=1}^n \mu_i^2 \\ &\geq \sum_{i=1}^n \lambda_i^2 - 2\sum_{i=1}^n \lambda_i\mu_i + \sum_{i=1}^n \mu_i^2 = \sum_{i=1}^n (\lambda_i - \mu_i)^2\end{aligned}$$

Because of the rank constraint at most k eigenvalues of $\mathbf{A}$ can be different from zero such that the minimum of the right-hand side is achieved by

$$\sum_{i=1}^n (\lambda_i-\mu_i)^2 = \sum_{i=1}^k (\lambda_i-\mu_i)^2 + \sum_{i=k+1}^n \lambda_i^2 \geq \sum_{i=1}^k (\lambda_i - \lambda_i^+)^2 + \sum_{i=k+1}^n \lambda_i^2\,.$$

Obviously, the minimum of (2.23) is attained at $\mathbf{A}^*$ in (2.24). ■

2.6 Orthogonal and Projection Matrices

In this chapter, we consider geometric properties of orthogonal and projection matrices. Such transformations play an important role when visualizing data in a two- or three-dimensional space, moreover, when determining a low-dimensional representation of high-dimensional data, as is done in Chap. 4. We will also see that projections can be constructed by selecting certain vectors out of an orthonormal system.

Orthogonal matrices describe rotations or reflections of vectors in a certain coordinate system without changing their length. We will first summarize some general properties of orthogonal matrices.

Lemma 2.31 *Let* $\mathbf{M} \in \mathbb{R}^{n \times n}$ *be an orthogonal matrix, i.e.,* $\mathbf{M}\mathbf{M}^T = \mathbf{I}_n$*. Then the following holds.*

(a) The product of orthogonal matrices is an orthogonal matrix.
(b) The absolute value of all eigenvalues of an orthogonal matrix is equal to one.
(c) $\|\mathbf{M}\mathbf{x}\|_2 = \|\mathbf{x}\|_2$ *for all* $\mathbf{x} \in \mathbb{R}^n$*, hence* $\mathbf{M}$ *is a linear* isometry. *Vice versa, if* $\|\mathbf{A}\mathbf{x}\|_2 = \|\mathbf{x}\|_2$ *for all* $\mathbf{x} \in \mathbb{R}^n$ *for some matrix* $\mathbf{A} \in \mathbb{R}^{n \times n}$*, then* $\mathbf{A}$ *is orthogonal.*
(d) $\det \mathbf{M} \in \{-1, 1\}$.

The proof of Lemma 2.31 is left to the reader, see Exercise 2.10.

Example 2.32 Consider a 2-dimensional orthogonal transformation by the matrix

$$\mathbf{M} = \mathbf{M}(\varphi) = \begin{pmatrix} \cos\varphi & -\sin\varphi \\ \sin\varphi & \cos\varphi \end{pmatrix} \tag{2.26}$$

with rotation angle $\varphi \in [0, 2\pi]$. Any vector $\mathbf{u} \in \mathbb{R}^2$ is rotated counterclockwise to $\mathbf{M}(\varphi)\mathbf{u}$ by angle φ around the origin.

Example 2.33 Consider a 3-dimensional rotation in Cartesian coordinates x, y, z. The matrix

$$\mathbf{M}_x = \mathbf{M}_x(\alpha) = \begin{pmatrix} 1 & 0 & 0 \\ 0 & \cos\alpha & -\sin\alpha \\ 0 & \sin\alpha & \cos\alpha \end{pmatrix} \tag{2.27}$$

rotates points in the y–z-plane by $\alpha \in [0, 2\pi]$ around the x axis. The matrix

$$\mathbf{M}_y = \mathbf{M}_y(\beta) = \begin{pmatrix} \cos\beta & 0 & \sin\beta \\ 0 & 1 & 0 \\ -\sin\beta & 0 & \cos\beta \end{pmatrix} \tag{2.28}$$

rotates the x–z-plane by $\beta \in [0, 2\pi]$ around the y axis, and the matrix

$$\mathbf{M}_z = \mathbf{M}_z(\gamma) = \begin{pmatrix} \cos\gamma & -\sin\gamma & 0 \\ \sin\gamma & \cos\gamma & 0 \\ 0 & 0 & 1 \end{pmatrix} \tag{2.29}$$

rotates the x–y-plane by $\gamma \in [0, 2\pi]$ around the z axis. All three matrices $\mathbf{M}_x$, $\mathbf{M}_y$ and $\mathbf{M}_z$ are orthogonal, and thus any product is again orthogonal. For example, the matrix

$$\mathbf{M}(\alpha, \beta, \gamma) = \mathbf{M}_x(\alpha)\,\mathbf{M}_y(\beta)\,\mathbf{M}_z(\gamma) \tag{2.30}$$

rotates a given point $\mathbf{u} \in \mathbb{R}^3$ first around the z-, then around the y- and finally around the x-axis. Since the multiplication of matrices is not commutative, interchanging the order of the product leads to a different result.

We next consider projection matrices as linear transformations onto a certain subspace. Typically, this subspace will have a small dimension. Projections are characterized by the property that once a point is projected onto a subspace there is no further change if the projection is applied once more. This is an obvious property of projections from a geometric point of view.

Definition 2.34 A matrix $\mathbf{Q} \in \mathbb{R}^{n\times n}$ is called *projection matrix* or *idempotent* if $\mathbf{Q}^2 = \mathbf{Q}$. It is called an *orthogonal projection* if additionally $\mathbf{Q}^T = \mathbf{Q}$.

The rank of a projection matrix $\mathbf{Q}$ determines the dimension of the hyperplane which it maps onto.

Projections may be skew w.r.t. the hyperplane they map onto. Symmetry of the projection matrix ensures that projections are orthogonal to the projection plane. This can be seen as follows.

Let $\mathbf{x} \in \mathbb{R}^n$, $\mathbf{x} \notin \text{Img}(\mathbf{Q})$ and $\mathbf{y} \in \text{Img}(\mathbf{Q})$. Then $(\mathbf{x} - \mathbf{Q}\mathbf{x})$ is perpendicular to $\mathbf{y}$ since $\mathbf{y}$ has a representation $\mathbf{y} = \mathbf{Q}\mathbf{z}$ for some $\mathbf{z} \in \mathbb{R}^n$ and

$$\mathbf{y}^T(\mathbf{x} - \mathbf{Q}\mathbf{x}) = \mathbf{z}^T\mathbf{Q}(\mathbf{x} - \mathbf{Q}\mathbf{x}) = \mathbf{z}^T(\mathbf{Q}\mathbf{x} - \mathbf{Q}\mathbf{x}) = 0\,.$$

This also shows that the distance between a point and its projected counterpart is minimal over all points in the projection hyperplane.

Note that orthogonal projections $\mathbf{Q}$ are nonnegative definite since for all $\mathbf{x} \in \mathbb{R}^n$

$$\mathbf{x}^T\mathbf{Q}\mathbf{x} = \mathbf{x}^T\mathbf{Q}^2\mathbf{x} = (\mathbf{Q}\mathbf{x})^T\mathbf{Q}\mathbf{x} = \mathbf{y}^T\mathbf{y} \geq 0\,. \tag{2.31}$$

In the sequel we will clarify how to construct projection matrices onto a certain subspace with a given orthonormal basis.

Lemma 2.35 *(a) Let* $k \leq n$ *and* $\mathbf{V} = (\mathbf{v}_1, \ldots, \mathbf{v}_k) \in \mathbb{R}^{n\times k}$ *such that* $\mathbf{V}^T\mathbf{V} = \mathbf{I}_k$. *Then*

$$\mathbf{Q} = \sum_{i=1}^{k} \mathbf{v}_i\mathbf{v}_i^T \tag{2.32}$$

is an orthogonal projection onto $\text{Img}(\mathbf{Q}) = \text{Span}\big(\{\mathbf{v}_1, \mathbf{v}_2, \ldots, \mathbf{v}_k\}\big)$.

(b) If $\mathbf{Q}$ *is an orthogonal projection onto* $\text{Img}(\mathbf{Q})$, *then* $\mathbf{I} - \mathbf{Q}$ *is an orthogonal projection onto* $\text{Ker}(\mathbf{Q})$.

Proof Obviously $\mathbf{Q}$ is symmetric and fulfills

$$\mathbf{Q}^2 = \sum_{i=1}^{k}\sum_{j=1}^{k} \mathbf{v}_i\mathbf{v}_i^T\mathbf{v}_j\mathbf{v}_j^T = \sum_{i=1}^{k} \mathbf{v}_i\mathbf{v}_i^T = \mathbf{Q}\,,$$

since $\mathbf{v}_i^T\mathbf{v}_j = 1$ if $i = j$, and 0 otherwise. Moreover, the image of an arbitrary point $\mathbf{x}$ under $\mathbf{Q}$ is given by

$$\mathbf{Q}\mathbf{x} = \sum_{i=1}^{k} \mathbf{v}_i \underbrace{\mathbf{v}_i^T \mathbf{x}}_{=\beta_i} = \sum_{i=1}^{k} \beta_i \mathbf{v}_i \in \operatorname{Span}\big(\{\mathbf{v}_1, \mathbf{v}_2, \ldots, \mathbf{v}_k\}\big),$$

which together with $\operatorname{rank}(\mathbf{Q}) = k$ shows that $\operatorname{Img}(\mathbf{Q}) = \operatorname{Span}(\{\mathbf{v}_1, \mathbf{v}_2, \ldots, \mathbf{v}_k\})$ and shows part (a).

Matrix $\mathbf{M} = \mathbf{I} - \mathbf{Q}$ is also an orthogonal projection since it is symmetric and fulfills

$$\mathbf{M}^2 = \mathbf{I} - 2\mathbf{Q} + \mathbf{Q}^2 = \mathbf{I} - 2\mathbf{Q} + \mathbf{Q} = \mathbf{M}.$$

If $\mathbf{y} \in \operatorname{Ker}(\mathbf{Q})$, i.e., $\mathbf{Q}\mathbf{y} = \mathbf{0}$, then $\mathbf{M}\mathbf{y} = \mathbf{y} - \mathbf{Q}\mathbf{y} = \mathbf{y}$ such that $\mathbf{y} \in \operatorname{Img}(\mathbf{M})$. On the other hand, each $\mathbf{x} \in \operatorname{Img}(\mathbf{M})$ has a representation as $\mathbf{x} = \mathbf{M}\mathbf{z}$ for some $\mathbf{z} \in \mathbb{R}^n$. Then $\mathbf{Q}\mathbf{x} = \mathbf{Q}\mathbf{M}\mathbf{z} = \big(\mathbf{Q} - \mathbf{Q}^2\big)\mathbf{z} = \mathbf{0}$ such that $\mathbf{x} \in \operatorname{Ker}(\mathbf{Q})$. In summary, we have $\operatorname{Img}(\mathbf{M}) = \operatorname{Ker}(\mathbf{Q})$, which completes the proof of part (b). ■

Example 2.36 The matrix $\mathbf{Q} = \frac{1}{n}\mathbf{1}_n\mathbf{1}_n^T = \frac{1}{n}\mathbf{1}_{n\times n}$ is an orthogonal projection onto $\operatorname{Img}(\mathbf{Q}) = \operatorname{Span}(\{\mathbf{1}_n\})$, the diagonal in $\mathbb{R}^n$. The only nonzero eigenvalue is $\lambda = 1$ with normalized eigenvector $\mathbf{v} = \frac{1}{\sqrt{n}}\mathbf{1}_n$. For any $\mathbf{x} = (x_1, \ldots, x_n)^T$ the product $\mathbf{Q}\mathbf{x}$ is the vector whose components are all equal to $\bar{x} = \frac{1}{n}\mathbf{1}^T\mathbf{x} = \frac{1}{n}\sum_i x_i$, the arithmetic mean of the components.

According to Lemma 2.35 the matrix

$$\mathbf{E}_n = \mathbf{I}_n - \frac{1}{n}\mathbf{1}_{n\times n} = \begin{pmatrix} 1-\frac{1}{n} & -\frac{1}{n} & -\frac{1}{n} & \cdots & -\frac{1}{n} \\ -\frac{1}{n} & 1-\frac{1}{n} & -\frac{1}{n} & \cdots & -\frac{1}{n} \\ -\frac{1}{n} & -\frac{1}{n} & 1-\frac{1}{n} & \cdots & -\frac{1}{n} \\ \vdots & \vdots & \vdots & \ddots & \vdots \\ -\frac{1}{n} & -\frac{1}{n} & -\frac{1}{n} & \cdots & 1-\frac{1}{n} \end{pmatrix}$$

is also an orthogonal projection. The image of $\mathbf{E}_n$ is the orthogonal complement of the diagonal in $\mathbb{R}^n$, namely $\operatorname{Img}(\mathbf{E}_n) = \{\mathbf{x} \in \mathbb{R}^n \mid \mathbf{1}_n^T\mathbf{x} = 0\}$. This subspace is also the kernel of $\mathbf{1}_{n\times n}$, since $\mathbf{1}_{n\times n}\mathbf{E}_n = \mathbf{1}_{n\times n} - \mathbf{1}_{n\times n} = \mathbf{0}_{n\times n}$.

Matrix $\mathbf{E}_n$ is called *centering matrix* since for any projected point $\mathbf{y} = \mathbf{E}_n\mathbf{x} = (y_1, \ldots, y_n)$ we have

$$\sum_{i=1}^{n} y_i = \mathbf{1}_n^T\mathbf{y} = \mathbf{1}_n^T\mathbf{E}_n\mathbf{x} = \mathbf{0}^T\mathbf{x} = 0.$$

This means that the arithmetic mean over the components of $\mathbf{y}$ is zero, it is hence centered at the origin.

2.7 Exercises

Exercise 2.1 Prove the following for a symmetric matrix $\mathbf{M}$. If $\mathbf{x}^T\mathbf{M}\mathbf{x} \geq 0$ for all $\mathbf{x} \in \mathbb{R}^n$, then $\mathbf{M}$ is nonnegative definite. If $\mathbf{x}^T\mathbf{M}\mathbf{x} > 0$ for all $\mathbf{x} \in \mathbb{R}^n$, $\mathbf{x} \neq \mathbf{0}$ then $\mathbf{M}$ is positive definite. This shows the reversed direction in Lemma 2.4 (c) and (d).

Exercise 2.2 Show that for all symmetric matrices $\mathbf{M}$ the Frobenius norm and the spectral norm satisfy $\|\mathbf{M}\|_\mathrm{F} \geq \|\mathbf{M}\|_\mathrm{S}$.

Exercise 2.3 Show properties (a)–(c) of the Frobenius and the spectral norm in Lemma 2.14.

Exercise 2.4 Show the properties (a)–(d) of the trace and its connections to eigenvalues and the Frobenius norm in Lemma 2.16.

Exercise 2.5 Use Gershgorin's Theorem to find the smallest area in the complex plane which contains all eigenvalues of the matrix $\mathbf{A}$. Is $\mathbf{A}$ positive definite? Determine the smallest interval $[a, b]$ which contains the real part of all eigenvalues.

$$\mathbf{A} = \begin{pmatrix} 10 & 0.1 & 1 & 0.9 & 0 \\ 0.2 & 9 & 0.2 & 0.2 & 0.2 \\ 0.3 & -0.1 & 5+i & 0 & 0.1 \\ 0 & 0.6 & 0.1 & 6 & -0.3 \\ 0.3 & -0.3 & 0.1 & 0 & 1 \end{pmatrix}$$

Exercise 2.6 Show that for a symmetric matrix $\mathbf{M}$ with eigenvalues $\lambda_1, \lambda_2, \ldots, \lambda_n$ the following holds:

$$\mathrm{tr}(\mathbf{M}^k) = \sum_{i=1}^{n} \lambda_i^k \quad \text{and} \quad \det(\mathbf{M}^k) = \prod_{i=1}^{n} \lambda_i^k$$

for all $k \in \mathbb{N}$.

Exercise 2.7 Show that for a matrix $\mathbf{M} = \begin{pmatrix} \alpha & \beta \\ \beta & \gamma \end{pmatrix}$ with entries $\alpha, \beta, \gamma \in \mathbb{R}$ the eigenvalues are given by

$$\lambda_{1/2}(\mathbf{M}) = \frac{\alpha + \gamma}{2} \pm \frac{1}{2}\sqrt{(\alpha - \gamma)^2 + 4\beta^2}\,.$$

Exercise 2.8 Consider a lever arm with a bearing at its left end and notches at certain distances $x_1, \ldots, x_n$ from the bearing, see Fig. 2.1. How would you hang weights $w_1, \ldots, w_n$ to the notches such that the torque is maximum or minimum? Is there a modification of the physical experiment so that negative distances and forces can be included.

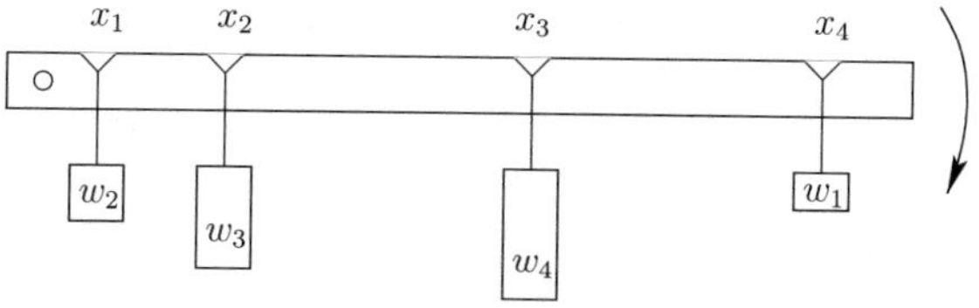

Fig. 2.1 Lever arm with a bearing on the left, notches at distances x_i and weights w_i

Exercise 2.9 Let $\mathbf{U} = (\mathbf{u}_1, \ldots \mathbf{u}_n)$ and $\mathbf{V} = (\mathbf{v}_1, \ldots \mathbf{v}_n)$ be orthogonal matrices. Show that

$$\sum_{i=1}^{n} \left(\mathbf{u}_i^T \mathbf{v}_j\right)^2 = \sum_{j=1}^{n} \left(\mathbf{u}_i^T \mathbf{v}_j\right)^2 = 1.$$

Exercise 2.10 Prove the properties given in Lemma 2.31.

Chapter 3
Multivariate Distributions and Moments

Probability theory provides mathematical laws for randomness and is hence an essential tool for quantitative analysis of nondeterministic or noisy data. It allows the description of complex systems when only partial knowledge of the state is available. For example, supervised learning is performed on the basis of training data. To assess robustness and reliability of derived decision and classification rules, knowledge of the underlying distributions is essential.

A central subject in the framework of data analytics are random vectors and multivariate distributions, which is the focus of this chapter.

3.1 Random Vectors

Foundations of modern probability theory were laid by Andrey Nikolaevich Kolmogorov, also known for his fundamental contributions to complexity theory. In 1933, he developed a system of axioms which forms the starting point of the rich theory available today.

We assume that the concept of a probability space $(\Omega, \mathcal{A}, P)$ is known to the reader. You should also be familiar with the definition of a real valued random variable (r.v.) $X : (\Omega, \mathcal{A}, P) \to (\mathbb{R}, \mathcal{B})$ as a mapping from Ω to $\mathbb{R}$ which additionally is measurable with respect to the Borel σ-algebra $\mathcal{B}$. By this, a random variable is a function which induces a certain probability distribution on its domain. The precise knowledge of the underlying probability space is mostly not needed. Its existence is necessary to deal with so-called *events* in a precise mathematical manner.

Now, let $X_1, \dots, X_p$ be real random variables on the same probability space

$$X_i : (\Omega, \mathcal{A}, P) \to (\mathbb{R}, \mathcal{B}), \quad i = 1, \dots, p.$$

The vector composed of the random variables

R. Mathar et al., *Fundamentals of Data Analytics*,
https://doi.org/10.1007/978-3-030-56831-3_3

$$X = (X_1, \ldots, X_p)^T : (\Omega, \mathcal{A}, P) \to (\mathbb{R}^p, \mathcal{B}^p)$$

is called *random vector*. Analogously, a *random matrix* is defined by combining random variables X_{ij}, $i = 1, \ldots, p$, $j = 1, \ldots, n$ into a matrix

$$X = \begin{pmatrix} X_{11} & \cdots & X_{1n} \\ \vdots & \ddots & \vdots \\ X_{p1} & \cdots & X_{pn} \end{pmatrix}.$$

The *joint probability distribution* of a random vector is uniquely defined by its *multivariate distribution function*

$$F(x_1, \ldots, x_p) = P(X_1 \le x_1, \ldots, X_p \le x_p), \quad (x_1, \ldots, x_p) \in \mathbb{R}^p.$$

A random vector is called *absolutely-continuous* if there exists an integrable function $f(x_1, \ldots, x_p) \ge 0$ such that

$$F(x_1, \ldots, x_p) = \int_{-\infty}^{x_1} \cdots \int_{-\infty}^{x_p} f(t_1, \ldots, t_p)\, \mathrm{d}t_1 \cdots \mathrm{d}t_p.$$

Function f is called *probability density function* (pdf) or just *density*, while F is called *cumulative distribution function* (CDF) or just *distribution function*.

Example 3.1 The multivariate normal or Gaussian distribution is defined by the density function

$$f(\mathbf{x}) = \frac{1}{(2\pi)^{p/2}|\boldsymbol{\Sigma}|^{1/2}} \exp\big\{-\tfrac{1}{2}(\mathbf{x}-\boldsymbol{\mu})^T \boldsymbol{\Sigma}^{-1}(\mathbf{x}-\boldsymbol{\mu})\big\}, \ \mathbf{x} = (x_1, \ldots, x_p)^T \in \mathbb{R}^p. \tag{3.1}$$

with parameters $\boldsymbol{\mu} \in \mathbb{R}^p$ and positive definite $\boldsymbol{\Sigma} \in \mathbb{R}^{p \times p}$. The notation will be $X = (X_1, \ldots, X_p)^T \sim N_p(\boldsymbol{\mu}, \boldsymbol{\Sigma})$.

Note that $\boldsymbol{\Sigma}$ must have full rank for a density to exist. If rank $\boldsymbol{\Sigma} < p$ we define a random vector X to be p-dimensional Gaussian distributed if $\mathbf{a}^T X$ is one-dimensional Gaussian or single-point distributed for all $\mathbf{a} \in \mathbb{R}$. Note that in case rank $\boldsymbol{\Sigma} < p$ there is no density with respect to the p-dimensional Lebesgue measure.

Figure 3.1 shows a graphical presentation of the two-dimensional Gaussian density for $\boldsymbol{\mu} = \mathbf{0}$ and different matrices $\boldsymbol{\Sigma}$.

3.2 Expectation and Covariance

The *expectation vector* or *mean vector*, and the *covariance matrix* will be derived in the following. We assume that the expected value and higher moments of a one-dimensional random variable are familiar concepts. In higher dimensions the mean

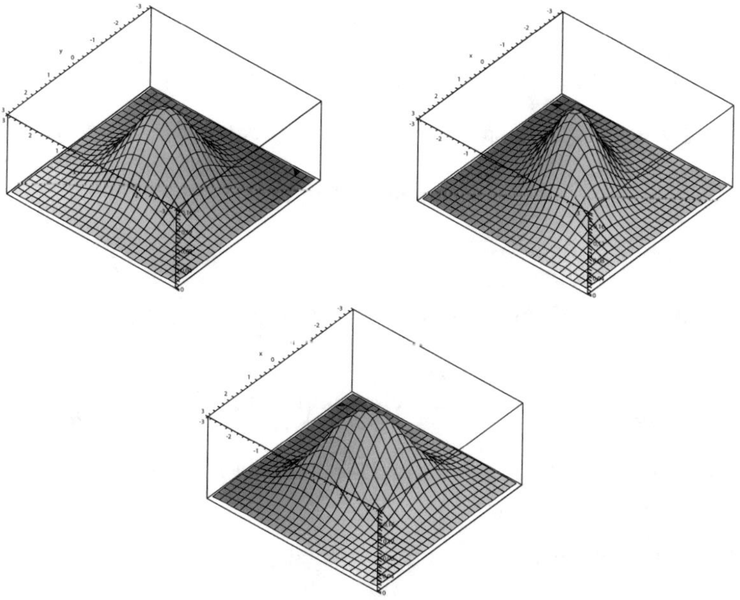

Fig. 3.1 Densities of two-dimensional Gaussians with $\boldsymbol{\mu} = \mathbf{0}$ and $\boldsymbol{\Sigma}_1 = \begin{pmatrix} 1 & 0 \\ 0 & 1 \end{pmatrix}$, $\boldsymbol{\Sigma}_2 = \begin{pmatrix} 1 & 0.5 \\ 0.5 & 1 \end{pmatrix}$, $\boldsymbol{\Sigma}_3 = \begin{pmatrix} 1 & -0.5 \\ -0.5 & 1 \end{pmatrix}$ (left, right, middle)

vector is an immediate generalization of the one-dimensional concept. On the diagonal of a covariance matrix we find the variance of the components of a random vector. Off-diagonal elements are the pairwise covariance between components.

In the following definitions we assume that the corresponding first and second moments exist.

Definition 3.2 For a random vector $\boldsymbol{X} = (X_1, \ldots, X_p)^T$

(a) $\mathrm{E}(\boldsymbol{X}) = \big(\mathrm{E}(X_1), \ldots, \mathrm{E}(X_p)\big)^T$ is called *expectation vector* or *mean vector* of $\boldsymbol{X}$.
(b) $\mathrm{Cov}(\boldsymbol{X}) = \mathrm{E}\big([\boldsymbol{X} - \mathrm{E}(\boldsymbol{X})][\boldsymbol{X} - \mathrm{E}(\boldsymbol{X})]^T\big)$ is called *covariance matrix* of $\boldsymbol{X}$.

Note that the expectation vector is simply composed of the componentwise expectations. The (i, j)-th entry of the $p \times p$ covariance matrix is the covariance between r.v. X_i and X_j, namely $\mathrm{E}\big([X_i - \mathrm{E}(X_i)][X_j - \mathrm{E}(X_j)]\big)$.

The following theorem summarizes some elementary rules for dealing with expectation vectors and covariance matrices. The proofs are left as an exercise.

Theorem 3.3 *Let $\boldsymbol{X}, \boldsymbol{Y} \in \mathbb{R}^p$ be random vectors, further $\mathbf{A} \in \mathbb{R}^{n \times p}$ a fixed matrix and $\mathbf{b} \in \mathbb{R}^p$ a fixed vector. Provided that the corresponding moments exist, we have*

(a) $\mathrm{E}(\mathbf{A}\boldsymbol{X} + \mathbf{b}) = \mathbf{A}\,\mathrm{E}(\boldsymbol{X}) + \mathbf{b}$,
(b) $\mathrm{E}(\boldsymbol{X} + \boldsymbol{Y}) = \mathrm{E}(\boldsymbol{X}) + \mathrm{E}(\boldsymbol{Y})$,
(c) $\mathrm{Cov}(\mathbf{A}\boldsymbol{X} + \mathbf{b}) = \mathbf{A}\,\mathrm{Cov}(\boldsymbol{X})\mathbf{A}^T$,

(d) $\mathrm{Cov}(X+Y) = \mathrm{Cov}(X) + \mathrm{Cov}(Y)$ *if* X *and* Y *are stochastically independent, and*

(e) $\mathrm{Cov}(X) \geq_{\mathrm{L}} 0$, *i.e.,* $\mathrm{Cov}(X)$ *is nonnegative definite.*

Example 3.4 If X is p-dimensional Gaussian distributed, $X \sim \mathcal{N}_p(\boldsymbol{\mu}, \boldsymbol{\Sigma})$, then

$$\mathrm{E}(X) = \boldsymbol{\mu} \text{ and } \mathrm{Cov}(X) = \boldsymbol{\Sigma},$$

see Exercise 3.2.

Theorem 3.5 (Steiner's rule) *For any random vector* X *and fixed vector* $\mathbf{b} \in \mathbb{R}^p$ *it holds that*

$$\mathrm{E}\big([X-b][X-b]^T\big) = \mathrm{Cov}(X) + \big(\mathbf{b} - \mathrm{E}(X)\big)\big(\mathbf{b} - \mathrm{E}(X)\big)^T. \tag{3.2}$$

Hence, the left hand side of equation (3.2) decomposes into the covariance of X *and an additive term called* bias.

Proof With the abbreviation $\boldsymbol{\mu} = \mathrm{E}(X)$ we obtain

$$\begin{aligned}
\mathrm{E}\big((X-b)(X-b)^T\big) &= \mathrm{E}\big((X-\boldsymbol{\mu}+\boldsymbol{\mu}-b)(X-\boldsymbol{\mu}+\boldsymbol{\mu}-b)^T\big) \\
&= \mathrm{E}\big((X-\boldsymbol{\mu})(X-\boldsymbol{\mu})^T\big) + \mathrm{E}\big((\boldsymbol{\mu}-b)(\boldsymbol{\mu}-b)^T\big) \\
&= \mathrm{Cov}(X) + (\boldsymbol{\mu}-b)(\boldsymbol{\mu}-b)^T,
\end{aligned}$$

where equality in the second line follows since $\mathrm{E}\big((X-\boldsymbol{\mu})(\boldsymbol{\mu}-b)^T\big) = \mathbf{0}_{p\times p}$ and $\mathrm{E}(\mathbf{b}) = \mathbf{b}$ for any $\mathbf{b} \in \mathbb{R}^p$. ■

With probability one any random vector realizes values within the image of its covariance matrix, shifted by the expectation. If the covariance matrix does not have full rank, the values are found on a low-dimensional hyperplane.

Theorem 3.6 *For any random vector* X *with* $\mathrm{E}(X) = \boldsymbol{\mu}$ *and* $\mathrm{Cov}(X) = \boldsymbol{\Sigma}$ *it holds that*

$$P\big(X \in \mathrm{Img}(\boldsymbol{\Sigma}) + \boldsymbol{\mu}\big) = 1.$$

Proof Recall that $\mathrm{Ker}(\boldsymbol{\Sigma}) = \{\mathbf{x} \in \mathbb{R}^p \mid \boldsymbol{\Sigma}\mathbf{x} = \mathbf{0}\}$ denotes the nullspace or kernel of $\boldsymbol{\Sigma}$ and let $\mathbf{a}_1, \dots, \mathbf{a}_r$ be a basis of $\mathrm{Ker}(\boldsymbol{\Sigma})$.

It holds that $\mathbf{a}_i^T\boldsymbol{\Sigma}\mathbf{a}_i = \mathrm{Var}(\mathbf{a}_i^T X) = 0$ for all $i = 1, \dots, r$, and hence

$$P\big(\mathbf{a}_i^T X = \mathrm{E}(\mathbf{a}_i^T X)\big) = P\big(\mathbf{a}_i^T X = \mathbf{a}_i^T \boldsymbol{\mu}\big) = P\big(\mathbf{a}_i^T (X - \boldsymbol{\mu}) = 0\big) = 1.$$

For the orthogonal subspaces $\mathbf{a}_i^\perp$ this means that $P\big(X - \boldsymbol{\mu} \in \mathbf{a}_i^\perp\big) = 1$ for all $i = 1, \dots, r$.

The relation (cf. Exercises 3.3 and 3.4)

$$\mathrm{Img}(\boldsymbol{\Sigma}) = \mathrm{Ker}(\boldsymbol{\Sigma})^\perp = \langle \mathbf{a}_1, \dots, \mathbf{a}_r \rangle^\perp = \mathbf{a}_1^\perp \cap \dots \cap \mathbf{a}_r^\perp$$

finally yields

$$P\big(X - \boldsymbol{\mu} \in \operatorname{Img}(\boldsymbol{\Sigma})\big) = P\big(X - \boldsymbol{\mu} \in \mathbf{a}_1^{\perp} \cap \cdots \cap \mathbf{a}_r^{\perp}\big) = 1,$$

which proves the assertion. ■

3.3 Conditional Distributions

Prior knowledge about the outcome of random experiments is modeled by conditional distributions. We deal with the case of absolutely-continuous random vectors with a joint density and assume that a part of the components is observable prior to defining the distribution of the remaining ones.

A given random vector $\boldsymbol{X} = (X_1, \ldots, X_p)^T$ with density $f_{\boldsymbol{X}}$ is hence split into two parts $\boldsymbol{X} = (\boldsymbol{Y}_1, \boldsymbol{Y}_2)$ with $\boldsymbol{Y}_1 = (X_1, \ldots, X_k)^T$ and $\boldsymbol{Y}_2 = (X_{k+1}, \ldots, X_p)^T$, $k < p$. Let $f_{\boldsymbol{Y}_2}(\mathbf{y}_2)$ denote the marginal density of the joint density $f_{\boldsymbol{Y}_1,\boldsymbol{Y}_2}$ for any $\mathbf{y}_2 \in \mathbb{R}^{p-k}$.

Definition 3.7

$$f_{\boldsymbol{Y}_1|\boldsymbol{Y}_2}(\mathbf{y}_1 \mid \mathbf{y}_2) = \frac{f_{\boldsymbol{Y}_1,\boldsymbol{Y}_2}(\mathbf{y}_1, \mathbf{y}_2)}{f_{\boldsymbol{Y}_2}(\mathbf{y}_2)}, \quad \mathbf{y}_1 \in \mathbb{R}^k$$

is called *conditional density* of $\boldsymbol{Y}_1$ given $\boldsymbol{Y}_2 = \mathbf{y}_2$.

The conditional density is hence obtained by dividing the joint density by the conditioning marginal one. It may happen that $f_{\boldsymbol{Y}_2}(\mathbf{y}_2) = 0$ for certain values of $\mathbf{y}_2$. In this case the conditional density is set to $f_{\boldsymbol{Y}_1}(\mathbf{y}_1)$.

A general property of conditional densities is the following. It may also be taken to almost surely define conditional densities in case a joint density exists, i.e.,

$$P(\boldsymbol{Y}_1 \in \mathcal{B} \mid \boldsymbol{Y}_2 = \mathbf{y}_2) = \int_{\mathcal{B}} f_{\boldsymbol{Y}_1|\boldsymbol{Y}_2}(\mathbf{y}_1, \mathbf{y}_2)\, \mathrm{d}\mathbf{y}_1$$

for any measurable set $\mathcal{B}$.

For jointly Gaussian distributed random vectors the conditional density is again Gaussian. Conditioning implies a change of parameters which will be considered next.

Theorem 3.8 *Suppose that* $(\boldsymbol{Y}_1, \boldsymbol{Y}_2) \sim N_p(\boldsymbol{\mu}, \boldsymbol{\Sigma})$ *with*

$$\boldsymbol{\mu} = \begin{pmatrix} \boldsymbol{\mu}_1 \\ \boldsymbol{\mu}_2 \end{pmatrix}, \quad \boldsymbol{\Sigma} = \begin{pmatrix} \boldsymbol{\Sigma}_{11} & \boldsymbol{\Sigma}_{12} \\ \boldsymbol{\Sigma}_{21} & \boldsymbol{\Sigma}_{22} \end{pmatrix} \quad \text{and} \quad \boldsymbol{\Sigma}^{-1} = \boldsymbol{\Lambda} = \begin{pmatrix} \boldsymbol{\Lambda}_{11} & \boldsymbol{\Lambda}_{12} \\ \boldsymbol{\Lambda}_{21} & \boldsymbol{\Lambda}_{22} \end{pmatrix}$$

of corresponding dimensions k *and* $p - k$ *with* $1 < k < p$. *Then the following holds:*

(a) $\boldsymbol{Y}_1$ *and* $\boldsymbol{Y}_2$ *are Gaussian distributed, namely*

$$\boldsymbol{Y}_1 \sim \mathcal{N}_k(\boldsymbol{\mu}_1, \boldsymbol{\Sigma}_{11}) \quad and \quad \boldsymbol{Y}_2 \sim \mathcal{N}_{p-k}(\boldsymbol{\mu}_2, \boldsymbol{\Sigma}_{22}) .$$

(b) *The conditional distribution of* $\boldsymbol{Y}_1$ *given* $\boldsymbol{Y}_2 = \mathbf{y}_2$ *is Gaussian with density*

$$f_{\boldsymbol{Y}_1|\boldsymbol{Y}_2}(\mathbf{y}_1 \mid \mathbf{y}_2) \sim \mathcal{N}_k(\boldsymbol{\mu}_{1|2}, \boldsymbol{\Sigma}_{1|2}) ,$$

where the parameters are given by

$$\boldsymbol{\mu}_{1|2} = \boldsymbol{\mu}_1 + \boldsymbol{\Sigma}_{12}\boldsymbol{\Sigma}_{22}^{-1}(\mathbf{y}_2 - \boldsymbol{\mu}_2) = \boldsymbol{\Sigma}_{1|2}\big(\boldsymbol{\Lambda}_{11}\boldsymbol{\mu}_1 - \boldsymbol{\Lambda}_{12}(\mathbf{y}_2 - \boldsymbol{\mu}_2)\big) \tag{3.3}$$

and by the so-called Schur-complementary matrix

$$\boldsymbol{\Sigma}_{1|2} = \boldsymbol{\Sigma}_{11} - \boldsymbol{\Sigma}_{12}\boldsymbol{\Sigma}_{22}^{-1}\boldsymbol{\Sigma}_{21} = \boldsymbol{\Lambda}_{11}^{-1}, \tag{3.4}$$

provided the inverse exists.

Part (b) of the theorem is proved by computing the quotient of the joint density and the marginal density of $\boldsymbol{Y}_2$. It requires careful calculus of the corresponding integrals and is left as an exercise.

3.4 Maximum Likelihood Estimation

We consider n independent samples from a certain distribution with density f depending on certain parameters θ. As an example, think of the multivariate Gaussian density (3.1) whose shape is determined by $\boldsymbol{\mu}$ and $\boldsymbol{\Sigma}$. More formally we have independent random samples $\mathbf{x}_1, \ldots, \mathbf{x}_n \in \mathbb{R}^p$ from a pdf $f(\mathbf{x}; \theta)$, where θ denotes a set of parameters. The joint density by which the observations are governed is the product of the same marginal density f. The goal now is to match the parameters to the observations in an optimal way. This is essentially the concept of maximum likelihood estimation (MLE). Let us start with some basic definitions.

Definition 3.9 For n observations $\mathbf{x}_1, \ldots, \mathbf{x}_n \in \mathbb{R}^p$ and a density $f(\mathbf{x}; \theta)$, $\mathbf{x} \in \mathbb{R}^p$, with parameter θ, the *likelihood function* is defined as

$$L(\mathbf{X}; \theta) = \prod_{i=1}^{n} f(\mathbf{x}_i; \theta),$$

while

$$\ell(\mathbf{X}; \theta) = \log L(\mathbf{X}; \theta) = \sum_{i=1}^{n} \log f(\mathbf{x}_i; \theta)$$

is called *log-likelihood function*, where we denote $\mathbf{X} = (\mathbf{x}_1, \ldots, \mathbf{x}_n)$.

Now consider a sample of independent random data points $\mathbf{x}_1, \ldots, \mathbf{x}_n \in \mathbb{R}^p$. By independence the joint pdf of the points is $\prod_{i=1}^n f(\mathbf{x}_i; \theta)$. The shape of density f is known, however, the parameter θ is unknown. The parameter θ which explains the given sample most likely is found by the following optimization problem.

Definition 3.10 Given $\mathbf{X} = (\mathbf{x}_1, \ldots, \mathbf{x}_n)$. The optimum value

$$\hat{\theta} = \arg\max_{\theta} \ell(\mathbf{X}; \theta)$$

is called *maximum likelihood estimator* (MLE).

Note that maximizing $\ell(\mathbf{X}; \theta)$ is equivalent to maximizing $L(\mathbf{X}; \theta)$ with the same solution, since the former is obtained by a monotone transformation of the latter.

Theorem 3.11 *Let $X \sim \mathcal{N}_p(\boldsymbol{\mu}, \boldsymbol{\Sigma})$ and $\mathbf{x}_1, \ldots, \mathbf{x}_n$ be independent samples from $\mathbf{X}$. The maximum likelihood estimators of $\boldsymbol{\mu}$ and $\boldsymbol{\Sigma}$ are*

$$\hat{\boldsymbol{\mu}} = \frac{1}{n}\sum_{i=1}^{n} \mathbf{x}_i = \bar{\mathbf{x}} \quad \text{and} \quad \widehat{\boldsymbol{\Sigma}} = \frac{1}{n}\sum_{i=1}^{n} (\mathbf{x}_i - \bar{\mathbf{x}})(\mathbf{x}_i - \bar{\mathbf{x}})^T = \mathbf{S}_n .$$

Proof The proof uses derivatives of real valued matrix functions, see A.4 in [8], or [24] for an in-depth treatment of the subject. It holds that

$$\frac{\partial}{\partial \mathbf{V}} \log |\mathbf{V}| = (\mathbf{V}^{-1})^T \tag{3.5}$$

and

$$\frac{\partial}{\partial \mathbf{V}} \operatorname{tr}(\mathbf{V}\mathbf{A}) = \mathbf{A}^T . \tag{3.6}$$

We furthermore use Steiner's rule (3.2). Let $\theta = (\boldsymbol{\mu}, \boldsymbol{\Sigma})$ and apply the natural logarithm on the density (3.1) of the multivariate Gaussian to obtain

$$\begin{aligned}
\ell\big((\mathbf{x}_1, \ldots, \mathbf{x}_n); \boldsymbol{\mu}, \boldsymbol{\Sigma}\big) &= \sum_{i=1}^{n}\Big[\log \frac{1}{(2\pi)^{p/2}|\boldsymbol{\Sigma}|^{1/2}} - \frac{1}{2}(\mathbf{x}_i - \boldsymbol{\mu})^T \boldsymbol{\Sigma}^{-1}(\mathbf{x}_i - \boldsymbol{\mu})\Big] \\
&= n \log \frac{1}{(2\pi)^{p/2}} + \frac{n}{2} \log |\boldsymbol{\Sigma}^{-1}| - \frac{1}{2}\sum_{i=1}^{n}(\mathbf{x}_i - \boldsymbol{\mu})^T \boldsymbol{\Sigma}^{-1}(\mathbf{x}_i - \boldsymbol{\mu}) .
\end{aligned}$$

The first term is constant w.r.t. maximization, so we can omit it and proceed with

$$\begin{aligned}
\ell'\big((\mathbf{x}_1,\ldots,\mathbf{x}_n);\boldsymbol{\mu},\boldsymbol{\Sigma}\big) &= \frac{n}{2}\log|\boldsymbol{\Sigma}^{-1}| - \frac{1}{2}\sum_{i=1}^{n}(\mathbf{x}_i-\boldsymbol{\mu})^T\boldsymbol{\Sigma}^{-1}(\mathbf{x}_i-\boldsymbol{\mu}) \\
&= \frac{n}{2}\log|\boldsymbol{\Sigma}^{-1}| - \frac{1}{2}\sum_{i=1}^{n}\operatorname{tr}\big(\boldsymbol{\Sigma}^{-1}(\mathbf{x}_i-\boldsymbol{\mu})(\mathbf{x}_i-\boldsymbol{\mu})^T\big) \\
&= \frac{n}{2}\log|\boldsymbol{\Sigma}^{-1}| - \frac{1}{2}\operatorname{tr}\Big(\boldsymbol{\Sigma}^{-1}\sum_{i=1}^{n}(\mathbf{x}_i-\boldsymbol{\mu})(\mathbf{x}_i-\boldsymbol{\mu})^T\Big) \\
&= \frac{n}{2}\log|\boldsymbol{\Sigma}^{-1}| - \frac{1}{2}\operatorname{tr}\Big(\boldsymbol{\Sigma}^{-1}\sum_{i=1}^{n}(\mathbf{x}_i-\bar{\mathbf{x}})(\mathbf{x}_i-\bar{\mathbf{x}})^T + (\bar{\mathbf{x}}-\boldsymbol{\mu})(\bar{\mathbf{x}}-\boldsymbol{\mu})^T\Big) \\
&\le \frac{n}{2}\log|\boldsymbol{\Sigma}^{-1}| - \frac{n}{2}\operatorname{tr}(\boldsymbol{\Sigma}^{-1}\mathbf{S}_n)
\end{aligned}$$

with $\mathbf{S}_n = \frac{1}{n}\sum_{i=1}^{n}(\mathbf{x}_i-\bar{\mathbf{x}})(\mathbf{x}_i-\bar{\mathbf{x}})^T$. Using the rules (3.5) and (3.6) with $\boldsymbol{\Phi} = \boldsymbol{\Sigma}^{-1}$ and setting the partial derivatives to zero yields

$$\frac{\partial}{\partial\boldsymbol{\Phi}}\Big(\frac{n}{2}\log|\boldsymbol{\Phi}| - \frac{n}{2}\operatorname{tr}\left(\boldsymbol{\Phi}\mathbf{S}_n\right)\Big) = \frac{n}{2}\boldsymbol{\Sigma} - \frac{n}{2}\mathbf{S}_n = \mathbf{0}.$$

Hence, the optimum solution is $\boldsymbol{\Sigma}^* = \mathbf{S}_n$. Equality in the above inequality is attained for $\boldsymbol{\mu} = \bar{\mathbf{x}}$, which completes the proof. ■

3.5 Exercises

Exercise 3.1 Prove the assertions of Theorem 3.3:

(a) $\mathrm{E}(\mathbf{A}X+\mathbf{b}) = \mathbf{A}\,\mathrm{E}(X)+\mathbf{b}$
(b) $\mathrm{E}(X+Y) = \mathrm{E}(X)+\mathrm{E}(Y)$
(c) $\mathrm{Cov}(\mathbf{A}X+\mathbf{b}) = \mathbf{A}\,\mathrm{Cov}(X)\mathbf{A}^T$
(d) $\mathrm{Cov}(X+Y) = \mathrm{Cov}(X)+\mathrm{Cov}(Y)$ if X and Y are stochastically independent
(e) $\mathrm{Cov}(X) \ge_{\mathrm{L}} 0$, i.e., $\mathrm{Cov}(X)$ is nonnegative definite

Exercise 3.2 Show that for a p-dimensional Gaussian distributed random vector $X \sim N_p(\boldsymbol{\mu},\boldsymbol{\Sigma})$ expectation and covariance are $\mathrm{E}(X) = \boldsymbol{\mu}$ and $\mathrm{Cov}(X) = \boldsymbol{\Sigma}$.

Exercise 3.3 Show that for a random vector X and (measurable) sets $\mathcal{A},\mathcal{B} \in \mathbb{R}^p$ the following holds:

$$P(X\in\mathcal{A}) = P(X\in\mathcal{B}) = 1 \Leftrightarrow P(X\in\mathcal{A}\cap\mathcal{B}) = 1\,.$$

Exercise 3.4 The orthogonal complement of the linear space spanned by vectors $\mathbf{a}_1,\ldots,\mathbf{a}_r$ coincides with the intersection of the orthogonal complement of each, i.e., $\langle\mathbf{a}_1,\ldots,\mathbf{a}_r\rangle^\perp = \mathbf{a}_1^\perp\cap\cdots\cap\mathbf{a}_r^\perp$. Why?

Exercise 3.5 Derive the formulas (3.3) and (3.4) for the expectation and covariance of the conditional distribution in Theorem 3.8 (b).

Exercise 3.6 Suppose that $(Y_1, Y_2) \sim \mathcal{N}_2(\boldsymbol{\mu}, \boldsymbol{\Sigma})$ with

$$\boldsymbol{\mu} = \begin{bmatrix} \mu_1 \\ \mu_2 \end{bmatrix}, \quad \boldsymbol{\Sigma} = \begin{bmatrix} \sigma_1^2 & \rho\sigma_1\sigma_2 \\ \rho\sigma_1\sigma_2 & \sigma_2^2 \end{bmatrix}$$

and $\mu_1, \mu_2 \in \mathbb{R}$, $\sigma_1, \sigma_2 \geq 0$, $|\rho| \leq 1$.

(a) For which values of σ_1, σ_2 and ρ is $\boldsymbol{\Sigma}$ positive definite?
(b) If $\boldsymbol{\Sigma}$ is positive definite, determine

- the joint density $f_{Y_1,Y_2}(y_1, y_2)$,
- the marginal densities of Y_1 and Y_2,
- the conditional density $f_{Y_1|Y_2}(y_1|y_2)$.

Exercise 3.7 Suppose that the random variable $X_1, \ldots, X_n$ are stochastically independent, each absolutely continuous with density

$$f_{X_i}(x; \lambda) = \begin{cases} \lambda e^{-\lambda x}, & \text{if } x > 0, \\ 0, & \text{if } x \leq 0, \end{cases}$$

with parameter $\lambda > 0$, the density of the so called *exponential distribution*. Determine the maximum likelihood estimator of λ from n observations $x_1, \ldots, x_n$ of r.v.s $X_1, \ldots, X_n$.

Chapter 4
Dimensionality Reduction

In many cases data analytics has to cope with the extremely high dimension of the input. Structures may be well hidden not only by the sheer amount of data but also by very high-dimensional noise added to relatively low-dimensional signals. The aim of this chapter is to introduce methods which represent high-dimensional data in a low-dimensional space in a way that only a minimum of core information is lost. Optimality will mostly refer to projections in Hilbert spaces. If dimension one, two or three is sufficient to represent the raw data, a computer aided graphical visualization may help to identify clusters or outlying objects.

Often Principal Component Analysis (PCA) is applied before further processing data because of two main reasons. Firstly, low-dimensional data or signals may be corrupted by high-dimensional noise. PCA would eliminate noise from the data and alleviate subsequent analysis. Secondly, some algorithms are inefficient in high dimensions, which prevents results in acceptable time. Prior dimension reduction by PCA can drastically speed up the process.

In this chapter, we will first deal with *Principal Component Analysis*, roughly spoken a method for reducing the dimension of data while retaining most of the underlying variance. Secondly, we will explain in detail *Multidimensional Scaling*, a method to construct an embedding of objects in a Euclidean space only from the knowledge of pairwise dissimilarities. The latter has some powerful applications in the area of data analytics.

4.1 Principal Component Analysis (PCA)

In general, Principal Component Analysis looks for a few linear combinations which represent the data in a way to lose as little information as possible. More specifically, for a given data set $\mathbf{x}_1, \ldots, \mathbf{x}_n \in \mathbb{R}^p$ the aim is

R. Mathar et al., *Fundamentals of Data Analytics*,
https://doi.org/10.1007/978-3-030-56831-3_4

- to find a k-dimensional subspace such that the projections of $\mathbf{x}_1, \ldots, \mathbf{x}_n$ thereon represent the original points on its best,
- or to find the k-dimensional projections that preserve as much variance as possible.

It will be shown that both objectives lead to the same result, namely the representation via the leading or principal eigenvalues and eigenvectors.

We assume that $\mathbf{x}_1, \ldots, \mathbf{x}_n \in \mathbb{R}^p$ are independently drawn from some distribution with existing covariance matrix. The sample mean is defined as

$$\bar{\mathbf{x}} = \frac{1}{n} \sum_{i=1}^{n} \mathbf{x}_i$$

and the sample covariance matrix

$$\mathbf{S}_n = \frac{1}{n-1} \sum_{i=1}^{n} (\mathbf{x}_i - \bar{\mathbf{x}})(\mathbf{x}_i - \bar{\mathbf{x}})^T.$$

The arithmetic mean $\bar{\mathbf{x}}$ is the maximum likelihood unbiased estimator of the expectation vector. Matrix $\mathbf{S}_n$ is an unbiased estimator of the underlying covariance matrix, deviating from the maximum likelihood estimator for the Gaussian distribution by a factor of $\frac{n-1}{n}$ only, see Theorem 3.11.

The MNIST data set may be considered as an example to get a feeling of a typical dimension and sample size. The number of pixels $p = 28 \times 28 = 784$ determines the dimensions and n may be as large as 50.000.

4.1.1 Finding the Optimum Projection

We commence by considering the following optimization problem

$$\underset{\substack{\mathbf{a} \in \mathbb{R}^p, \\ \mathbf{Q} \in \mathbb{R}^{p \times p} \text{symmetric}, \mathbf{Q}^2 = \mathbf{Q}, \text{rank}(\mathbf{Q}) = k}}{\text{minimize}} \quad \sum_{i=1}^{n} \left\| (\mathbf{x}_i - \mathbf{a}) - \mathbf{Q}(\mathbf{x}_i - \mathbf{a}) \right\|^2. \tag{4.1}$$

The conditions $\mathbf{Q} \in \mathbb{R}^{p \times p}$ symmetric, $\mathbf{Q}^2 = \mathbf{Q}$, $\text{rank}(\mathbf{Q}) = k$ determine the set of orthogonal projections onto a k-dimensional subspace. The idea behind this problem is to find a shift vector $\mathbf{a}$ and an orthogonal projection such that the projected points are closest to the original ones.

The following transformations reduce the problem to one whose solution is known by Ky Fan's Theorem 2.27. The variables are $\mathbf{a} \in \mathbb{R}^p$ and $\mathbf{Q} \in \mathbb{R}^{p \times p}$ an orthogonal projection with $\text{rank}(\mathbf{Q}) = k$.

$$\min_{\mathbf{a},\mathbf{Q}} \sum_{i=1}^{n} \|\mathbf{x}_i - \mathbf{a} - \mathbf{Q}(\mathbf{x}_i - \mathbf{a})\|^2 = \min_{\mathbf{a},\mathbf{Q}} \sum_{i=1}^{n} \|(\mathbf{I} - \mathbf{Q})(\mathbf{x}_i - \mathbf{a})\|^2$$
$$= \min_{\mathbf{a},\mathbf{R}} \sum_{i=1}^{n} \|(\mathbf{R}(\mathbf{x}_i - \mathbf{a})\|^2 ,$$

where $\mathbf{R} = \mathbf{I} - \mathbf{Q}$ is an orthogonal projection as well. Rewriting the squared norm as inner product and keeping in mind that $\mathrm{tr}(\mathbf{AB}) = \mathrm{tr}(\mathbf{BA})$, if the dimensions of matrices $\mathbf{A}$ and $\mathbf{B}$ are appropriate, we obtain

$$\begin{aligned}
\min_{\mathbf{a},\mathbf{R}} \sum_{i=1}^{n} (\mathbf{x}_i - \mathbf{a})^T \mathbf{R}^T \mathbf{R}(\mathbf{x}_i - \mathbf{a}) &= \min_{\mathbf{a},\mathbf{R}} \sum_{i=1}^{n} \mathrm{tr}\left((\mathbf{x}_i - \mathbf{a})^T \mathbf{R}(\mathbf{x}_i - \mathbf{a})\right) \\
&= \min_{\mathbf{a},\mathbf{R}} \sum_{i=1}^{n} \mathrm{tr}\left(\mathbf{R}(\mathbf{x}_i - \mathbf{a})(\mathbf{x}_i - \mathbf{a})^T\right) \\
&= \min_{\mathbf{a},\mathbf{R}} \mathrm{tr}\left(\mathbf{R} \sum_{i=1}^{n} (\mathbf{x}_i - \mathbf{a})(\mathbf{x}_i - \mathbf{a})^T\right) \\
&\geq \min_{\mathbf{R}} \mathrm{tr}\left(\mathbf{R} \sum_{i=1}^{n} (\mathbf{x}_i - \bar{\mathbf{x}})(\mathbf{x}_i - \bar{\mathbf{x}})^T\right) \\
&= \min_{\mathbf{R}} \mathrm{tr}\left(\mathbf{R}(n-1)\mathbf{S}_n\right) \\
&= \min_{\mathbf{Q}} (n-1)\, \mathrm{tr}\left(\mathbf{S}_n(\mathbf{I} - \mathbf{Q})\right).
\end{aligned}$$

It remains to solve the following optimization problem over all orthogonal projection matrices $\mathbf{Q} \in \mathbb{R}^{p\times p}$ of rank k. Each $\mathbf{Q}$ has a representation $\mathbf{Q} = \sum_{i=1}^{k} \mathbf{q}_i \mathbf{q}_i^T$ with $\mathbf{q}_i$ pairwise orthonormal so that $\mathbf{Q} = \tilde{\mathbf{Q}}\tilde{\mathbf{Q}}^T$ with $\tilde{\mathbf{Q}} = (\mathbf{q}_1, \ldots, \mathbf{q}_k)$. Hence by Theorem 2.27

$$\max_{\mathbf{Q}} \mathrm{tr}\left(\mathbf{S}_n \mathbf{Q}\right) = \max_{\tilde{\mathbf{Q}}^T\tilde{\mathbf{Q}} = \mathbf{I}_k} \mathrm{tr}\left(\tilde{\mathbf{Q}}^T \mathbf{S}_n \tilde{\mathbf{Q}}\right) = \sum_{i=1}^{k} \lambda_i(\mathbf{S}_n), \tag{4.2}$$

where $\lambda_1(\mathbf{S}_n) \geq \cdots \geq \lambda_p(\mathbf{S}_n)$ are the eigenvalues of $\mathbf{S}_n$ in decreasing order. The maximum is attained if $\mathbf{q}_1, \ldots, \mathbf{q}_k$ are the orthonormal eigenvectors corresponding to $\lambda_1(\mathbf{S}_n), \ldots, \lambda_k(\mathbf{S}_n)$.

The projection matrix attaining the minimum in (4.1) is hence given by

$$\mathbf{Q}^* = \tilde{\mathbf{Q}}\tilde{\mathbf{Q}}^T = \sum_{i=1}^{k} \mathbf{q}_i \mathbf{q}_i^T \in \mathbb{R}^{p\times p} \tag{4.3}$$

with the orthonormal eigenvectors $\mathbf{q}_i$ corresponding to the k largest eigenvalues of the sample covariance matrix $\mathbf{S}_n$.

Having determined the k most significant eigenvectors $\mathbf{q}_1, \ldots, \mathbf{q}_k$, each data $\mathbf{x}_i$ is projected onto

$$\hat{\mathbf{x}}_i = \sum_{\ell=1}^{k} \mathbf{q}_\ell \mathbf{q}_\ell^T \mathbf{x}_i = \sum_{\ell=1}^{k} \mathbf{q}_\ell^T \mathbf{x}_i \mathbf{q}_\ell.$$

The vectors $\left(\mathbf{q}_1^T \mathbf{x}_i, \ldots, \mathbf{q}_k^T \mathbf{x}_i\right), i = 1, \ldots, n$, are hence the representation in the linear space spanned by the first k unit vectors, for $k \leq 3$ appropriate for a low-dimensional visualization.

4.1.2 Preserving Most Variance

Another reasonable objective when reducing the dimension is to find an orthogonal projection matrix $\mathbf{Q}$ such that the projected points retain the maximum amount of variance in the data. This problem reads as

$$\begin{aligned}
\max_{\mathbf{Q}} \sum_{i=1}^{n} \left\| \mathbf{Q}\mathbf{x}_i - \frac{1}{n} \sum_{\ell=1}^{n} \mathbf{Q}\mathbf{x}_\ell \right\|^2 &= \max_{\mathbf{Q}} \sum_{i=1}^{n} \|\mathbf{Q}(\mathbf{x}_i - \bar{\mathbf{x}})\|^2 \\
&= \max_{\mathbf{Q}} \sum_{i=1}^{n} \operatorname{tr}\left((\mathbf{x}_i - \bar{\mathbf{x}})^T \mathbf{Q}(\mathbf{x}_i - \bar{\mathbf{x}})\right) \\
&= \max_{\mathbf{Q}} \operatorname{tr}\left(\mathbf{Q} \sum_{i=1}^{n} (\mathbf{x}_i - \bar{\mathbf{x}})(\mathbf{x}_i - \bar{\mathbf{x}})^T\right) \\
&= \max_{\mathbf{Q}} (n-1) \operatorname{tr}\left(\mathbf{Q}\mathbf{S}_n\right)
\end{aligned}$$

over all orthogonal projection matrices $\mathbf{Q} \in \mathbb{R}^{p \times p}$ having a representation $\mathbf{Q} = \tilde{\mathbf{Q}}\tilde{\mathbf{Q}}^T$ with $\tilde{\mathbf{Q}} \in \mathbb{R}^{p \times k}$ such that $\tilde{\mathbf{Q}}^T \tilde{\mathbf{Q}} = \mathbf{I}_k$. Obviously, the maximum is attained at solution (4.3), so that both approaches in Sects. 4.1.1 and 4.1.2 are equivalent.

The ratio

$$\rho_n = \frac{\sum_{i=1}^{n} \|\mathbf{Q}\mathbf{x}_i - \frac{1}{n}\sum_{\ell=1}^{n} \mathbf{Q}\mathbf{x}_\ell\|^2}{\sum_{i=1}^{n} \|\mathbf{x}_i - \bar{\mathbf{x}}\|^2} \tag{4.4}$$

measures the percentage of variance which is explained by the projected points $\mathbf{Q}\mathbf{x}_i$. For the optimum projection $\mathbf{Q}^*$ from (4.3) it follows from the above that

$$
\begin{aligned}
\sum_{i=1}^{n} \left\| \mathbf{Q}^*\mathbf{x}_i - \tfrac{1}{n}\textstyle\sum_{\ell=1}^{n} \mathbf{Q}^*\mathbf{x}_\ell \right\|^2 &= (n-1)\,\mathrm{tr}\left(\mathbf{Q}^*\mathbf{S}_n\right) \\
&= (n-1)\,\mathrm{tr}\left(\sum_{\ell=1}^{k} \lambda_\ell \mathbf{q}_\ell \mathbf{q}_\ell^T\right) \\
&= (n-1)\sum_{\ell=1}^{k} \lambda_\ell\,,
\end{aligned}
$$

since $\mathbf{Q}^* = \sum_{\ell=1}^{k} \mathbf{q}_\ell \mathbf{q}_\ell^T$ with orthonormal eigenvectors $\mathbf{q}_\ell$ corresponding to the k largest eigenvectors λ_ℓ of $\mathbf{S}_n$. Similarly we conclude that

$$
\sum_{i=1}^{n} \|\mathbf{x}_i - \bar{\mathbf{x}}\|^2 = (n-1)\sum_{\ell=1}^{p} \lambda_\ell.
$$

Ratio (4.4) becomes

$$
\rho_n = \frac{\sum_{\ell=1}^{k} \lambda_\ell}{\sum_{\ell=1}^{p} \lambda_\ell} \tag{4.5}
$$

measuring the percentage of variance retained by the PCA projection onto a k-dimensional subspace in terms of the k largest eigenvalues of the sample covariance matrix $\mathbf{S}_n$.

4.1.3 *Finding the Right k*

Ratio (4.5) is the percentage of preserved variance of the projected points where $\lambda_1 \geq \cdots \geq \lambda_k$ are the eigenvalues of $\mathbf{S}_n$ in decreasing order. We would exclude smaller values of λ_ℓ from the projection since only a small amount of variance is removed by the corresponding projection.

An easy method to determine the right k is to observe the *scree plot* of the ordered eigenvalues [9]. A drastic drop in the curve, a so called *elbow* indicates that the remaining variance is small compared to the one determined by the previous principal eigenvalues. A typical example is shown in Fig. 4.1, in that case $k = 3$ would clearly be chosen.

4.1.4 *Computational Aspects of PCA*

The following steps are carried out to find a representation of some data set $\mathbf{x}_1, \ldots \mathbf{x}_n \in \mathbb{R}^p$ in a low dimensional space.

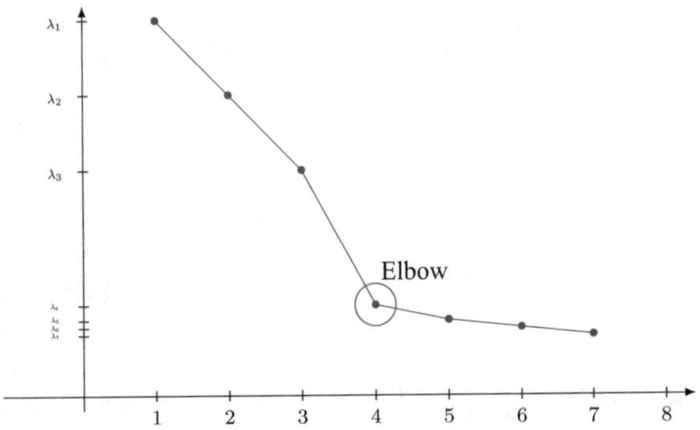

Fig. 4.1 An example of a scree plot where an essential part of the variance is combined in the first three principal eigenvalues

1. Compute $\bar{\mathbf{x}} = \frac{1}{n}\sum_{i=1}^{n}\mathbf{x}_i$ and $\mathbf{S}_n = \frac{1}{n-1}\sum_{i=1}^{n}(\mathbf{x}_i - \bar{\mathbf{x}})(\mathbf{x}_i - \bar{\mathbf{x}})^T$.
2. Compute the spectral decomposition $\mathbf{S}_n = \mathbf{Q}\boldsymbol{\Lambda}\mathbf{Q}^T$,
   $\boldsymbol{\Lambda} = \operatorname{diag}(\lambda_1, \ldots \lambda_p)$, $\lambda_1 \geq \cdots \geq \lambda_p$, $\mathbf{Q} = (\mathbf{q}_1, \ldots, \mathbf{q}_p)$ orthogonal.
3. Fix k, the dimension of the low-dimensional space.
4. The k-dimensional representation of $\mathbf{x}_i$ is

$$\hat{\mathbf{x}}_i = \begin{pmatrix} \mathbf{q}_1^T \\ \vdots \\ \mathbf{q}_k^T \end{pmatrix} (\mathbf{x}_i - \bar{\mathbf{x}}) \in \mathbb{R}^k, \quad i = 1, \ldots, n.$$

The eigenvectors $\mathbf{q}_1, \ldots, \mathbf{q}_k$ are called the k *principle eigenvectors* to the *principle eigenvalues* $\lambda_1, \ldots, \lambda_k$.

The computational complexity of the above standard method requires $O(np^2)$ steps to construct the empirical covariance matrix $\mathbf{S}_n$ and $O(p^3)$ steps for the spectral decomposition. In total there are $O(\max\{np^2, p^3\})$ steps needed to determine the principal components.

Since not the full spectrum of eigenvalues and eigenvectors is needed but only the k largest eigenvalues and corresponding eigenvectors, there are methods, cf. Theorem 2.9, to reduce the computational complexity of PCA.

If $p < n$, which can be assumed in most cases of big data analysis, we can do better as follows. First we compose the data matrix $\mathbf{X} = [\mathbf{x}_1, \ldots, \mathbf{x}_n]$. The empirical covariance matrix or sample covariance matrix is then obtained as

$$\mathbf{S}_n = \frac{1}{n-1}(\mathbf{X} - \bar{\mathbf{x}}\mathbf{1}_n^T)(\mathbf{X} - \bar{\mathbf{x}}\mathbf{1}_n^T)^T.$$

The singular value decomposition of $\mathbf{X} - \bar{\mathbf{x}}\mathbf{1}_n^T \in \mathbb{R}^{p \times n}$ is given by

$$\mathbf{X} - \bar{\mathbf{x}}\mathbf{1}_n^T = \mathbf{U}\operatorname{diag}(\sigma_1, \ldots, \sigma_p)\mathbf{V}^T, \tag{4.6}$$

with $\mathbf{U} \in \mathbb{R}^{p \times p}$ orthogonal and $\mathbf{V} \in \mathbb{R}^{p \times n}$ such that $\mathbf{V}^T\mathbf{V} = \mathbf{I}_p$. Let $\mathbf{D} = \text{diag}(\sigma_1, \dots, \sigma_p)$ be the diagonal matrix of singular values. Then

$$\mathbf{S}_n = \frac{1}{n-1}\mathbf{U}\mathbf{D}\mathbf{V}^T\mathbf{V}\mathbf{D}\mathbf{U}^T = \frac{1}{n-1}\mathbf{U}\mathbf{D}^2\mathbf{U}^T \tag{4.7}$$

such that $\mathbf{U} = [\mathbf{u}_1, \dots, \mathbf{u}_p]$ contains the eigenvectors of $\mathbf{S}_n$.

The computational complexity of (4.6) is $O(\min\{n^2p, p^2n\})$. If one is only interested in the top k eigenvectors then the complexity even reduces to $O(knp)$.

The book [20] (also available on the author's web page) deals with efficient algorithms for the spectral decomposition of matrices with a view to big data applications. A major part is on randomized algorithms for computing the SVD of massive matrices and corresponding error bounds.

4.1.5 The Eigenvalue Structure of $\mathbf{S}_n$ in High Dimensions

We now suppose that the observations $\mathbf{x}_1, \dots, \mathbf{x}_n$ are independent realizations of a Gaussian random vector $X \sim \mathcal{N}_p(\mathbf{0}, \mathbf{\Sigma})$ with expectation $\mathbf{0}$ and covariance matrix $\mathbf{\Sigma}$. The data matrix $\mathbf{X} = [\mathbf{x}_1, \dots, \mathbf{x}_n] \in \mathbb{R}^{p \times n}$ is column-wise composed of the samples. The sample covariance matrix S_n is the maximum likelihood estimator of $\mathbf{\Sigma}$, see Theorem 3.11, and can be written as

$$S_n = \frac{1}{n}\sum_{i=1}^{n}\mathbf{x}_i\mathbf{x}_i^T = \frac{1}{n}\mathbf{X}\mathbf{X}^T.$$

Clearly, S_n is a random matrix so that the corresponding eigenvalues and eigenvectors are random variables and vectors as well.

If the dimension p is fixed, it follows from the strong law of large numbers that $\mathbf{S}_n \to \mathbf{\Sigma}$ almost surely as $n \to \infty$. However, if both n and p become large it is not anymore immediate what the relation between S_n and $\mathbf{\Sigma}$ is.

We will investigate the case that $\mathbf{\Sigma} = \mathbf{I}_n$, i.e., uncorrelated components in each sample. Since all eigenvalues of $\mathbf{I}_n$ are unit, we would expect that the distribution of eigenvalues of S_n for large n and p groups around 1. This is not true as can be seen from the simulation results depicted in Fig. 4.2. Quite the opposite, eigenvalues are spread over an interval of width ca. 2.8.

The dimension is $p = 500$, and $n = 1000$ random samples are generated from a Gaussian distribution $\mathcal{N}_p(\mathbf{0}, \mathbf{I}_p)$. The upper graph shows the scree plot of eigenvalues, the lower one depicts the histogram of eigenvalues. One can see that the eigenvalues range approximately between 0.1 and 2.9 and follow a certain distribution, whose density is indicated by a blue line. The scree plot also shows that there is no low-dimensional space to adequately represent the data by applying PCA.

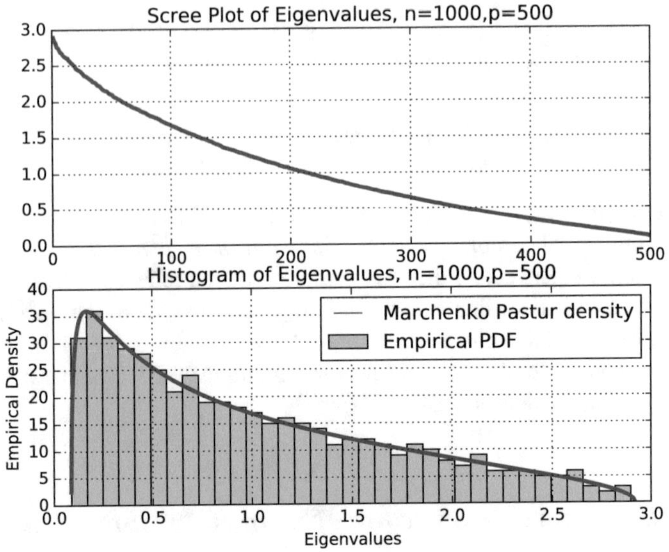

Fig. 4.2 Scree plot and histogram of the eigenvalues of the empirical covariance matrix $\mathbf{S}_n$ for 1000 samples from a $\mathcal{N}_{500}(\mathbf{0}, \mathbf{I})$ distribution

The following theorem provides a solid ground for the above observations. The eigenvalue distribution of $\boldsymbol{S}_n$ is determined as the number of observations n tends to infinity and the dimension p grows proportionally with n.

Theorem 4.1 (Marčenko-Pastur, 1967) *Let $\boldsymbol{X}_1, \ldots, \boldsymbol{X}_n \in \mathbb{R}^p$ be i.i.d. random vectors with $\mathrm{E}(\boldsymbol{X}_i) = \mathbf{0}$ and $\mathrm{Cov}(\boldsymbol{X}_i) = \sigma^2 \mathbf{I}_p$. Let $\boldsymbol{X} = (\boldsymbol{X}_1, \ldots, \boldsymbol{X}_n) \in \mathbb{R}^{p \times n}$ and $\boldsymbol{S}_n = \frac{1}{n}\boldsymbol{X}\boldsymbol{X}^T \in \mathbb{R}^{p \times p}$. Let $\lambda_1, \ldots, \lambda_p$ be the eigenvalues of $\boldsymbol{S}_n$. Suppose that $p, n \to \infty$ such that $\frac{p}{n} \to \gamma \in (0, 1]$ as $n \to \infty$. Then the sample distribution of $\lambda_1, \ldots, \lambda_p$ converges almost surely to the following density*

$$f_\gamma(u) = \begin{cases} \frac{1}{2\pi\sigma^2 u\gamma}\sqrt{(b-u)(u-a)}, & \text{if } a(\gamma) \le u \le b(\gamma), \\ 0, & \text{otherwise}, \end{cases} \tag{4.8}$$

with $a(\gamma) = \sigma^2(1 - \sqrt{\gamma})^2$ and $b(\gamma) = \sigma^2(1 + \sqrt{\gamma})^2$.

The proof of the Marčenko-Pastur theorem [25] is omitted here. A comprehensive proof also for the case of non-identically distributed columns can be found in [2, pp. 47–57].

The blue curve in Fig. 4.2 depicts this density for $\gamma = 0.5$ with $a(\gamma) = 0.086$ and $b(\gamma) = 2.914$.

Remark 4.2 If $\gamma > 1$, there will be a mass point at zero with probability $1 - 1/\gamma$. If $\gamma > 1$, then $n < p$ for sufficiently large n and p. Moreover, the rank of $\boldsymbol{S}_n = \frac{1}{n}\boldsymbol{X}\boldsymbol{X}^T$ will be at most $\min(p, n)$ which is n in this case. Hence, $\boldsymbol{S}_n \in \mathbb{R}^{p \times p}$ does not have full rank and zero is an eigenvalue with multiplicity of at least $p - n$ for large p and n.

Theorem 4.1 shows that spectrum of eigenvalues even in the case of i.i.d. distributed random variables is rather widespread. An important question is to what degree PCA can recover a low-dimensional structure from data subject to high-dimensional additive noise. This aspect is investigated in the next section.

4.1.6 A Spike Model

We model the situation that a low-dimensional signal is disturbed by high-dimensional noise. Observations are drawn from i.i.d. random vectors $\boldsymbol{X}_1, \ldots, \boldsymbol{X}_n \in \mathbb{R}^p$. Each $\boldsymbol{X}_i$ is assumed to be composed of a one-dimensional random signal $\sqrt{\beta} V_i \mathbf{v}$ with a fixed vector $\mathbf{v} \in \mathbb{R}$, $\|\mathbf{v}\| = 1$, signal intensity $\beta > 0$ and additionally additive p-dimensional noise $\boldsymbol{U}_i$. We hence start from the model

$$\boldsymbol{X}_i = \boldsymbol{U}_i + \sqrt{\beta} V_i \mathbf{v}, \quad i = 1, \ldots, n, \tag{4.9}$$

with the assumptions

$$\mathrm{E}(\boldsymbol{U}_i) = \mathbf{0}, \; \mathrm{Cov}(\boldsymbol{U}_i) = \mathbf{I}_p, \; \mathrm{E}(V_i) = 0, \; \mathrm{Var}(V_i) = 1.$$

Random variables $\boldsymbol{U}_i$ and V_i are assumed to be stochastically independent so that by Theorem 3.3 (c) and (d)

$$\mathrm{Cov}(\boldsymbol{X}_i) = \mathrm{Cov}(\boldsymbol{U}_i) + \mathbf{v}\,\mathrm{Var}(\sqrt{\beta} V_i)\mathbf{v}^T = \mathbf{I}_p + \beta \mathbf{v}\mathbf{v}^T.$$

Parameter β controls the average power of the signal.

We first observe empirically what happens for $\beta = 0.5$ and $\mathbf{v} = \mathbf{e}_1$ in the case of noise dimension $p = 500$ and $n = 1000$ observations generated from the corresponding distributions. The eigenvalue histogram of $\mathbf{S}_n$ is depicted in the upper graph of Fig. 4.3. Obviously no eigenvalue sticks out, the situation is very similar to seeing merely noise with no signal present. The conclusion is that PCA will not detect the one-dimensional signal hidden by noise, the signal-to-noise ratio is too low.

The situation changes if $\beta = 1.5$ is chosen with the other parameters unaltered. In the lower graph of Fig. 4.3 one eigenvalue prominently sticks out with a value of about 3.3. Applying PCA would yield a single dominant eigenvalue well isolated from the others, so that the direction $\mathbf{v}$ of the one-dimensional signal would be detected.

The question arises if for increasing p and n such that the quotient p/n tends to a constant, would there be some bound such that for β less than the bound the largest eigenvalue will not be distinguishable, but otherwise it would. The answer is given by the following theorem [3].

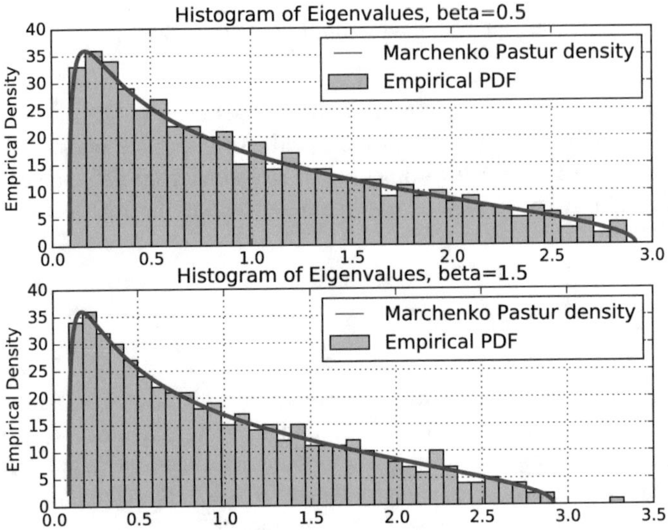

Fig. 4.3 Eigenvalue histogram of $\mathbf{S}_n$ for the spike model with $\beta = 0.5$ (upper graph) and $\beta = 1.5$ (lower graph). Other parameters are $p = 500,\ n = 1000$, $\mathbf{v} = \mathbf{e}_1$

Theorem 4.3 *Let $\boldsymbol{X}_1, \ldots, \boldsymbol{X}_n \in \mathbb{R}^p$ be i.i.d. random vectors with* $\mathrm{E}(\boldsymbol{X}_i) = \mathbf{0}$ *and* $\mathrm{Cov}(\boldsymbol{X}_i) = \mathbf{I}_p + \beta \mathbf{v}\mathbf{v}^T,\ \beta > 0,\ \mathbf{v} \in \mathbb{R}^p,\ \|\mathbf{v}\| = 1$. *Let* $\boldsymbol{X} = (\boldsymbol{X}_1, \ldots, \boldsymbol{X}_n) \in \mathbb{R}^{p \times n}$ *and* $\boldsymbol{S}_n = \frac{1}{n}\boldsymbol{X}\boldsymbol{X}^T \in \mathbb{R}^{p \times p}$. *Suppose that* $p, n \to \infty$ *such that* $\frac{p}{n} \to \gamma \in (0, 1]$ *as* $n \to \infty$. *Then the following holds almost surely:*

(a) *If* $\beta \le \sqrt{\gamma}$ *then* $\lambda_{\max} \to (1 + \sqrt{\gamma})^2$ *and* $|\langle \boldsymbol{v}_{\max}, \mathbf{v}\rangle|^2 \to 0$.
(b) *If* $\beta > \sqrt{\gamma}$ *then* $\lambda_{\max} \to (1 + \beta)(1 + \frac{\gamma}{\beta}) > (1 + \sqrt{\gamma})^2$ *and* $|\langle \boldsymbol{v}_{\max}, \mathbf{v}\rangle|^2 \to \frac{1-\gamma/\beta^2}{1-\gamma/\beta}$.

Herein, $\lambda_{\max}$ *and* $\boldsymbol{v}_{\max}$ *denote the largest eigenvalue and corresponding eigenvector of* $\boldsymbol{S}_n$.

The proof is omitted, it needs methods not provided in the present book. A good reference dealing not only with the largest but also with the ν-th largest eigenvalue is [33].

Recall from Theorem 4.1 that for $\sigma^2 = 1$ the right endpoint of the support of the Marčenko-Pasture distribution is $(1 + \sqrt{\gamma})^2$. Hence, in the spike model only if $\beta > \sqrt{\gamma}$ the largest eigenvalue of $\boldsymbol{S}_n$ exceeds the support of the Marčenko-Pasture distribution asymptotically, and hence is detectable by PCA. The corresponding density is also depicted in Fig. 4.3.

Furthermore, if β increases, then $\frac{1-\gamma/\beta^2}{1-\gamma/\beta} \to 1$ as $\beta \to \infty$, so that $\boldsymbol{v}_{\max}$ and $\mathbf{v}$ have an inner product of 1 and hence become aligned. In other words, the signal direction is detected by PCA. Vice versa, if $\beta \le \sqrt{\gamma}$, then $\boldsymbol{v}_{\max}$ and $\mathbf{v}$ are asymptotically perpendicular.

4.2 Multidimensional Scaling

Consider n points $\mathbf{x}_1, \ldots, \mathbf{x}_n$ in a p-dimensional Euclidean space and the corresponding data matrix $\mathbf{X} = [\mathbf{x}_1, \ldots, \mathbf{x}_n] \in \mathbb{R}^{p \times n}$, also called *configuration*. The distance between $\mathbf{x}_i$ and $\mathbf{x}_j$ is measured by the Euclidean norm and denoted as

$$d_{ij}(\mathbf{X}) = \|\mathbf{x}_i - \mathbf{x}_j\| = \sqrt{\sum_{\ell=1}^{p} \left(x_{\ell i} - x_{\ell j}\right)^2},$$

where $\mathbf{x}_i = (x_{1i} \ldots, x_{pi})^T$ and $\mathbf{x}_j = (x_{1j} \ldots, x_{pj})^T$. The pairwise distances are combined in a so called $n \times n$ *distance matrix*

$$\mathbf{D}(\mathbf{X}) = \left(d_{ij}(\mathbf{X})\right)_{i,j=1,\ldots,n} \in \mathbb{R}^{n \times n}.$$

Because of the properties of the above norm, $\mathbf{D}(\mathbf{X})$ is symmetric, has nonnegative entries and all diagonal elements are equal to 0.

Deriving $\mathbf{D}(\mathbf{X})$ from the configuration $\mathbf{X}$ is simple and immediate. Now consider the reversed problem. Given an $n \times n$ symmetric matrix $\mathbf{\Delta} = \left(\delta_{ij}\right)_{i,j=1,\ldots,n}$ with nonnegative entries and 0s on the diagonal. Is there a configuration $\mathbf{X} = [\mathbf{x}_1, \ldots, \mathbf{x}_n] \in \mathbb{R}^{p \times n}$ such that $\mathbf{\Delta} = \mathbf{D}(\mathbf{X})$? This is one of the key questions of Multidimensional Scaling (MDS), cf. [40]. A more detailed question will be to find the minimum dimension d such that a configuration in a d-dimensional Euclidean space exists. We call such a configuration *Euclidean embedding* of matrix $\mathbf{\Delta}$.

If there is no Euclidean embedding, an important problem is to find the best approximating configuration in a low-dimensional space, hence reducing dimensionality of the data. Solving this problem is highly relevant for certain applications of data analytics.

It is clear that if there exists one configuration, then there are infinitely many, since any configuration obtained from $\mathbf{X}$ by an orthogonal transformation and shift has the same distance matrix.

Let us start with a few simple examples. Given the matrix

$$\mathbf{\Delta} = \begin{pmatrix} 0 & 1 & 1 \\ 1 & 0 & 1 \\ 1 & 1 & 0 \end{pmatrix},$$

we ask whether there is a configuration in $\mathbb{R}^2$ which has $\mathbf{\Delta}$ as its distance matrix. The answer is yes. The corner points of any isosceles triangle of side length 1 is a corresponding configuration.

Now consider the matrix

$$\mathbf{\Delta} = \begin{pmatrix} 0 & 1 & 1 & 1 \\ 1 & 0 & 1 & 1 \\ 1 & 1 & 0 & 1 \\ 1 & 1 & 1 & 0 \end{pmatrix}.$$

Is there a configuration of four points in $\mathbb{R}^3$ with $\mathbf{\Delta}$ as distance matrix? Yes there is, any tetrahedron of side length 1 has $\mathbf{\Delta}$ as its distance matrix. Is there also a configuration in $\mathbb{R}^2$? You may try for a while and get the feeling there is none. But exactly why is this the case? In the next section we will comprehensively clarify these questions.

Before proceeding we fix the general notation. Consider n objects/points in $\mathbb{R}^p$. The pairwise *dissimilarities* δ_{ij} between object i and j are given. Symmetry and nonnegativity is assumed, i.e., $\delta_{ij} = \delta_{ji} \geq 0$, furthermore, diagonal elements are assumed to be zero. Matrix $\mathbf{\Delta} = \big(\delta_{ij}\big)_{1 \leq i,j \leq n}$ is called a *dissimilarity matrix*. The set of all dissimilarity matrices of dimension n is defined as

$$\mathcal{U}_n = \big\{\mathbf{\Delta} = (\delta_{ij})_{1\leq i,j \leq n} \mid \delta_{ij} = \delta_{ji} \geq 0, \delta_{ii} = 0\big\}. \tag{4.10}$$

We further introduce element-wise exponentiation of matrices with exponent $q > 0$ as

$$\mathbf{D}^{(q)}(\mathbf{X}) = \big(d_{ij}^q(\mathbf{X})\big)_{1\leq i,j\leq n} \text{ and } \mathbf{\Delta}^{(q)} = \big(\delta_{ij}^q\big)_{1\leq i,j \leq n}.$$

A general approximation problem in MDS is as follows. Given a dissimilarity matrix $\mathbf{\Delta} \in \mathcal{U}$, further some exponent $q \in \mathbb{N}$ and dimension $k \in \mathbb{N}$, solve the problem

$$\text{minimize}_{\mathbf{X} \in \mathbb{R}^{p \times n}} \ \big\|\mathbf{\Delta}^{(q)} - \mathbf{D}^{(q)}(\mathbf{X})\big\| \tag{4.11}$$

for some matrix-norm $\|\cdot\|$. In words, we want to find a configuration $\mathbf{X}$ of n points in a k-dimensional Euclidean space such that the corresponding distance matrix $\mathbf{D}^{(q)}(\mathbf{X})$ approximates the given dissimilarity matrix $\mathbf{\Delta}^{(q)}$ at its best. We hereby represent the dissimilarities by a cloud of n points in $\mathbb{R}^k$, which would allow for further processing, e.g., a graphical presentation if $k \leq 3$.

4.2.1 *Characterizing Euclidean Distance Matrices*

A dissimilarity matrix $\mathbf{\Delta} = (\delta_{ij})_{1\leq i,j\leq n} \in \mathcal{U}_n$ is called *Euclidean distance matrix* or is said to have a *Euclidean embedding* in $\mathbb{R}^k$ if there are $\mathbf{x}_1, \ldots, \mathbf{x}_n \in \mathbb{R}^k$ such that $\delta_{ij}^2 = \|\mathbf{x}_i - \mathbf{x}_j\|^2$. A dissimilarity matrix has a Euclidean embedding in k dimensions if approximation problem (4.11) with $q = 2$ has a solution with zero residuum.

Let

$$\mathbf{E}_n = \mathbf{I}_n - \tfrac{1}{n}\mathbf{1}_n\mathbf{1}_n^T = \mathbf{I}_n - \tfrac{1}{n}\mathbf{1}_{n\times n} \tag{4.12}$$

be the projection matrix onto the orthogonal complement of the diagonal in $\mathbb{R}^n$.

We start with a Lemma characterizing the null space of dissimilarity matrices which are doubly centered by $\mathbf{E}_n$.

Lemma 4.4 *Let* $\mathbf{A} \in \mathcal{U}_n$. *If* $\mathbf{E}_n\mathbf{A}\mathbf{E}_n = \mathbf{0}_{n\times n}$ *then* $\mathbf{A} = \mathbf{0}_{n\times n}$.

Proof The (i, j)-th element of $\mathbf{E}_n\mathbf{A}\mathbf{E}_n$ is easily computed to be

$$\left(\mathbf{E}_n\mathbf{A}\mathbf{E}_n\right)_{ij} = a_{ij} - a_{i\cdot} - a_{\cdot j} + a_{\cdot\cdot} = 0, \tag{4.13}$$

by assumption. The notation $a_{i\cdot}$, $a_{\cdot j}$ and $a_{\cdot\cdot}$ means summation over the dotted indices, e.g., $a_{i\cdot} = \sum_{\ell=1}^{n} a_{i\ell}$.

Because of symmetry $a_{i\cdot} = a_{\cdot i}$ holds. From $a_{ii} - a_{i\cdot} - a_{\cdot i} + a_{\cdot\cdot} = 0$ and $a_{ii} = 0$ it follows that $a_{i\cdot} = a_{\cdot\cdot}/2$. Analogously we conclude $a_{\cdot j} = a_{\cdot\cdot}/2$. From (4.13) it finally follows that $a_{ij} = 0$ for all $i, j = 1, \dots, n$, and thus $\left(\mathbf{E}_n\mathbf{A}\mathbf{E}_n\right) = \mathbf{0}_{n\times n}$. ■

The initial version of the following theorem goes back to a paper by Schoenberg from 1935, see [36].

Theorem 4.5 (Schoenberg, 1935) *A given dissimilarity matrix* $\boldsymbol{\Delta} \in \mathcal{U}_n$ *has a Euclidean embedding in k dimensions if and only if* $-\frac{1}{2}\mathbf{E}_n\boldsymbol{\Delta}^{(2)}\mathbf{E}_n$ *is nonnegative definite and* $\operatorname{rank}(\mathbf{E}_n\boldsymbol{\Delta}^{(2)}\mathbf{E}_n) \le k$. *The least* $k \in \mathbb{N}$ *which allows for an embedding is called the dimensionality of* $\boldsymbol{\Delta}$.

Proof For a given configuration $\mathbf{X} = [\mathbf{x}_1, \dots, \mathbf{x}_n] \in \mathbb{R}^{k\times n}$ it holds that

$$-\frac{1}{2}\mathbf{D}^{(2)}(\mathbf{X}) = \mathbf{X}^T\mathbf{X} - \mathbf{1}_n\hat{\mathbf{x}}^T - \hat{\mathbf{x}}\,\mathbf{1}_n^T, \tag{4.14}$$

where $\hat{\mathbf{x}} = \frac{1}{2}[\mathbf{x}_1^T\mathbf{x}_1, \dots, \mathbf{x}_n^T\mathbf{x}_n]^T$. Using this relation and the fact that $\mathbf{E}_n\mathbf{1}_n = \mathbf{0}$, we get

$$-\frac{1}{2}\mathbf{E}_n\mathbf{D}^{(2)}(\mathbf{X})\mathbf{E}_n = \mathbf{E}_n\mathbf{X}^T\mathbf{X}\,\mathbf{E}_n \ge_{\mathrm{L}} 0$$

and $\operatorname{rank}\left(\mathbf{E}_n\mathbf{X}^T\mathbf{X}\,\mathbf{E}_n\right) \le k$, since $\mathbf{X} \in \mathbb{R}^{k\times n}$.

Now assume that there is a Euclidean embedding, i.e., $\boldsymbol{\Delta}^{(2)} = \mathbf{D}^{(2)}(\mathbf{X})$ for some configuration $\mathbf{X} \in \mathbb{R}^{k\times n}$. Then by (4.14)

$$-\frac{1}{2}\mathbf{E}_n\boldsymbol{\Delta}^{(2)}(\mathbf{X})\mathbf{E}_n = -\frac{1}{2}\mathbf{E}_n\mathbf{D}^{(2)}(\mathbf{X})\mathbf{E}_n = \mathbf{E}_n\mathbf{X}^T\mathbf{X}\mathbf{E}_n \ge_{\mathrm{L}} 0$$

is nonnegative definite and $\operatorname{rank}\left(\mathbf{E}_n\boldsymbol{\Delta}^{(2)}(\mathbf{X})\mathbf{E}_n\right) \le k$.

On the other hand let $-\frac{1}{2}\mathbf{E}_n\boldsymbol{\Delta}^{(2)}(\mathbf{X})\mathbf{E}_n \ge_{\mathrm{L}} 0$ be nonnegative definite and $\operatorname{rank}\left(\mathbf{E}_n\boldsymbol{\Delta}^{(2)}(\mathbf{X})\mathbf{E}_n\right) \le k$. Then by the spectral decomposition of $-\frac{1}{2}\mathbf{E}_n\boldsymbol{\Delta}^{(2)}(\mathbf{X})\mathbf{E}_n$ there exists some $\mathbf{X} \in \mathbb{R}^{k\times n}$ such that

$$-\frac{1}{2}\mathbf{E}_n\boldsymbol{\Delta}^{(2)}(\mathbf{X})\mathbf{E}_n = \mathbf{X}^T\mathbf{X} \text{ and } \mathbf{X}\mathbf{E}_n = \mathbf{X}.$$

Matrix $\mathbf{X} = [\mathbf{x}_1, \ldots, \mathbf{x}_n]$ is an appropriate configuration, since by (4.14)

$$-\frac{1}{2}\mathbf{E}_n\mathbf{D}^{(2)}(\mathbf{X})\mathbf{E}_n = \mathbf{E}_n\mathbf{X}^T\mathbf{X}\mathbf{E}_n = \mathbf{X}^T\mathbf{X} = -\frac{1}{2}\mathbf{E}_n\boldsymbol{\Delta}^{(2)}\mathbf{E}_n.$$

Because both $d_{ii}(\mathbf{X}) = 0$ and $\delta_{ii} = 0$, it follows from Lemma 4.4 that $\boldsymbol{\Delta}^{(2)} = \mathbf{D}^{(2)}(\mathbf{X})$. ■

4.2.2 *The Best Euclidean Fit to a Given Dissimilarity Matrix*

In the following the distance between matrices is measured by the *Frobenius norm*

$$\|\mathbf{A}\|_{\mathrm{F}} = \Big(\sum_{i,j} a_{ij}^2\Big)^{1/2}$$

as introduced in Sect. 2.2. It is a direct extension of the Euclidean norm to the linear space of matrices and fulfills the general axioms of a norm which are: (i) $\|\mathbf{A}\|_{\mathrm{F}} \geq 0$ and $\|\mathbf{A}\|_{\mathrm{F}} = 0$ if and only if $\mathbf{A} = \mathbf{0}$, (ii) $\|\alpha\mathbf{A}\|_{\mathrm{F}} = |\alpha|\,\|\mathbf{A}\|_{\mathrm{F}}$ for all $\alpha \in \mathbb{R}$, and (iii) the triangle inequality $\|\mathbf{A}+\mathbf{B}\|_{\mathrm{F}} \leq \|\mathbf{A}\|_{\mathrm{F}} + \|\mathbf{B}\|_{\mathrm{F}}$.

If there is no Euclidean embedding for a given dissimilarity matrix in a certain dimension k, we want to find the best approximating configuration $\mathbf{X} \in \mathbb{R}^{k\times n}$. In the set of dissimilarity matrices $\mathcal{U}_n$ optimality is measured by the norm

$$\|\mathbf{E}_n\mathbf{A}\mathbf{E}_n\|_{\mathrm{F}}, \;\; \mathbf{A} \in \mathcal{U}_n.$$

This is indeed a norm on the linear space $\mathcal{U}_n$, as can be seen by checking the corresponding axioms and using Lemma 4.4, see Exercise 4.6. Using this norm, the following theorem determines the best approximating configuration by minimizing the deviation between squared dissimilarities and squared distances.

Recall that the *positive part* of some real number λ is defined as $\lambda^+ = \max\{\lambda, 0\}$. It simply truncates negative values at zero.

Theorem 4.6 *Given* $\boldsymbol{\Delta} \in \mathcal{U}_n$. *Let*

$$-\frac{1}{2}\mathbf{E}_n\boldsymbol{\Delta}^{(2)}\mathbf{E}_n = \mathbf{V}\,\mathrm{diag}(\lambda_1, \ldots, \lambda_n)\mathbf{V}^T$$

be the spectral decomposition with eigenvalues $\lambda_1 \geq \cdots \geq \lambda_n$ *and corresponding eigenvectors* $\mathbf{V} = [\mathbf{v}_1, \ldots, \mathbf{v}_n]$. *Then*

$$\mathrm{minimize}_{\mathbf{X}\in\mathbb{R}^{n\times k}} \;\; \big\|\mathbf{E}_n\big(\boldsymbol{\Delta}^{(2)} - \mathbf{D}^{(2)}(\mathbf{X})\big)\mathbf{E}_n\big\|_F$$

has a solution

$$\mathbf{X}^* = \left[\sqrt{\lambda_1^+}\,\mathbf{v}_1, \ldots, \sqrt{\lambda_k^+}\,\mathbf{v}_k\right]^T \in \mathbb{R}^{k \times n}.$$

Proof The solution of

$$\text{minimize}_{\mathbf{A} \geq_L 0,\, \text{rank}(\mathbf{A}) \leq k} \ \left\| -\tfrac{1}{2}\mathbf{E}_n \boldsymbol{\Delta}^{(2)} \mathbf{E}_n - \mathbf{A} \right\|_F$$

is known from Theorem 2.30 to be

$$\mathbf{A}^* = \mathbf{V}\,\text{diag}(\lambda_1^+, \ldots, \lambda_k^+, 0, \ldots, 0)\mathbf{V}^T.$$

Using (4.14) it follows that

$$\begin{aligned}
-\frac{1}{2}\mathbf{E}_n \mathbf{D}^{(2)}(\mathbf{X}^*)\mathbf{E}_n &= \mathbf{E}_n {\mathbf{X}^*}^T \mathbf{X}^* \mathbf{E}_n \\
&= \mathbf{E}_n[\mathbf{v}_1, \ldots, \mathbf{v}_k]\,\text{diag}(\lambda_1^+, \ldots, \lambda_k^+)[\mathbf{v}_1, \ldots, \mathbf{v}_k]^T \mathbf{E}_n \\
&= \mathbf{V}\,\text{diag}(\lambda_1^+, \ldots, \lambda_k^+, 0, \ldots, 0)\mathbf{V}^T = \mathbf{A}^*,
\end{aligned}$$

so that the minimum is attained in the set $\left\{ -\frac{1}{2}\mathbf{E}_n \mathbf{D}^{(2)}(\mathbf{X})\mathbf{E}_n \mid \mathbf{X} \in \mathbb{R}^{k \times n} \right\}$. ■

Approximations of dissimilarity matrices with respect to other norms, corresponding algorithms, and other typical applications of MDS are considered in the lecture notes [28].

4.3 Nonlinear Dimensionality Reduction

Suppose that the data points are elements of a low-dimensional nonlinear manifold. A typical example is shown in Fig. 4.4 with points embedded on a 'Swiss roll'. After unrolling the manifold, MDS would exhibit the essentially two-dimensional structure. A direct two-dimensional embedding by MDS would not reveal this structure. Short Euclidean distances between points on overlaying parts of the surface would make them look close, although they are far apart following a path on the manifold.

The key idea to overcome this problem and reveal the low-dimensional structure is approximating the (unknown) *geodesic distances* on the manifold by weighted edges of a certain graph, and apply MDS considering the approximations as dissimilarities.

4.3.1 *Isometric Feature Mapping*

The *isometric feature mapping* (IsoMap) is a method for non-linear dimensionality reduction, see [4, 39]. Given data $\mathbf{x}_1, \ldots, \mathbf{x}_n \in \mathbb{R}^p$ lying on a hidden manifold, e.g., the Swiss roll, the idea is to approximate the geodesic distance of the data points

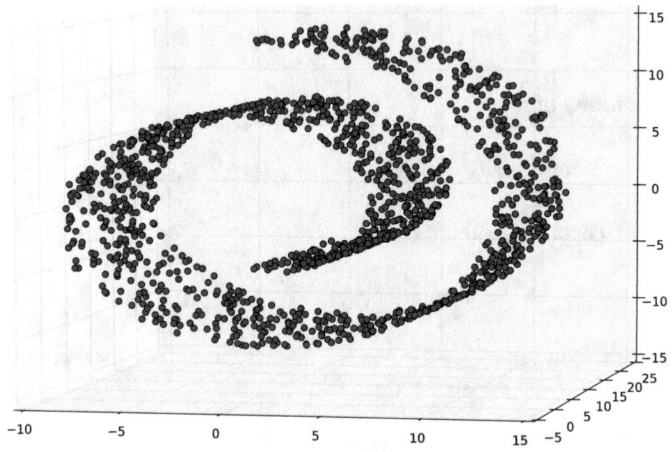

Fig. 4.4 'Swiss roll' data consisting of 1.500 random samples

by constructing a weighted graph and finding the shortest path between vertices. In a weighted graph each edge between vertices is endowed with a real number, the weight. Determining appropriate weights is an essential part of the IsoMap. The algorithm consists of three steps:

1. Given a real number $\epsilon > 0$. Construct a *neighborhood graph* $G = (V, E, \mathbf{W})$ with vertices $v_i = \mathbf{x}_i$ such that two vertices v_i and v_j are connected by an edge if $\|\mathbf{x}_i - \mathbf{x}_j\| \leq \epsilon$.
2. For each pair (v_i, v_j) compute the shortest path by Dijkstra's algorithm, cf. [37, Chap. 4.4]. The geodesic distance $\delta(v_i, v_j)$ is approximated by the number of hops or links from v_i to v_j. Alternatively the sum of all $\|\mathbf{x}_l - \mathbf{x}_k\|$ on a shortest path may serve as $\delta(v_i, v_j)$. Note that the geodesic distance can locally be well approximated by the Euclidean distance.
3. Construct a k-dimensional Euclidean embedding by applying MDS with dissimilarities $\Delta = \big(\delta(v_i, v_j)\big)_{1 \leq i,j \leq n}$.

4.3.2 Diffusion Maps

Diffusion Maps is a nonlinear technique for dimensionality reduction or feature extraction. Data is represented in a low-dimensional Euclidean space by using the diffusion distance generated by a related random walk. The main intention is to discover some underlying manifold that the data has been sampled from, see [10].

We start from contemplating a weighted graph $G = (V, E, \mathbf{W})$. Since the data will be embedded in a Euclidean space based on distances, it is reasonable to assume that the weight matrix $\mathbf{W} = (w_{ij})_{1 \leq i,j \leq n}$ is symmetric. Nodes which are connected by an edge with a large weight are considered to be close.

A homogeneous random walk X_t, $t = 0, 1, 2, \ldots$ is constructed on the graph with transition matrix

$$\mathbf{M} = (m_{ij})_{i,j=1,\ldots,n} \text{ with } m_{ij} = \frac{w_{ij}}{\deg(v_i)}, \quad 1 \leq i, j \leq n. \tag{4.15}$$

Here the *degree* of node v_i is denoted by $\deg(v_i) = \sum_{\ell=1}^{n} w_{i\ell}$.

Hence, the one-step transition probabilities of the random walk are

$$P(X_t = v_j \mid X_{t-1} = v_i) = m_{ij}, \; t = 1, 2, \ldots.$$

The conditional distribution of visiting node v_j after t steps having started in node v_i is given by the j-th entry of the i-th row of $\mathbf{M}^t = (m_{ij}^{(t)})_{1 \leq i,j \leq n}$, i.e.,

$$P(X_t = v_j \mid X_0 = v_i) = m_{ij}^{(t)}, \; j = 1, \ldots, n.$$

In vector form the t-step transition probabilities are obtained as the product of the i-th unit vector $\mathbf{e}_i$ with $\mathbf{M}^t$ as

$$\mathbf{u}_i = \mathrm{e}_i^T \mathbf{M}^t = (m_{i1}^{(t)}, \ldots, m_{in}^{(t)}).$$

If nodes v_i and v_j are close, expressed by a high transition probability, then the conditional distributions $\mathbf{u}_i$ and $\mathbf{u}_j$ will be similar. Starting in node v_i or starting in node v_j, connected by a high transition probability, will not make a big difference in the probability of dwelling in the same small cloud of points after t steps of the random walk. Theorem 4.10 will state this intuition more precisely. The random walk can be considered as a model for a diffusion process, which leads to the same intuition and also gives the technique its name.

In the following a smart representation of the distribution $\mathbf{u}_i$, $i = 1, \ldots, n$, is derived which allows for an embedding of the data in a low-dimensional space. The starting point is a bi-spectral decomposition of $\mathbf{M}^t$.

The transition matrix can be written as

$$\mathbf{M} = \mathbf{D}^{-1}\mathbf{W}, \quad \mathbf{D} = \mathrm{diag}\big(\deg(v_1), \ldots \deg(v_n)\big).$$

In general, $\mathbf{M} = \mathbf{D}^{-1}\mathbf{W}$ is not symmetric, however the normalized matrix

$$\mathbf{S} = \mathbf{D}^{1/2}\mathbf{M}\,\mathbf{D}^{-1/2} \tag{4.16}$$

is, since $\mathbf{D}^{1/2}\mathbf{D}^{-1}\mathbf{W}\mathbf{D}^{-1/2} = \mathbf{D}^{-1/2}\mathbf{W}\mathbf{D}^{-1/2}$ and $\mathbf{W}$ is symmetric. Matrix $\mathbf{S}$ has a spectral decomposition

$$\mathbf{S} = \mathbf{V}\boldsymbol{\Lambda}\mathbf{V}^T, \; \boldsymbol{\Lambda} = \mathrm{diag}(\lambda_1, \ldots, \lambda_n) \text{ with } \lambda_1 \geq \cdots \geq \lambda_n.$$

Hence

$$\mathbf{M} = \mathbf{D}^{-1/2}\mathbf{S}\mathbf{D}^{1/2} = \mathbf{D}^{-1/2}\mathbf{V}\boldsymbol{\Lambda}\mathbf{V}^T\mathbf{D}^{1/2} = (\mathbf{D}^{-1/2}\mathbf{V})\boldsymbol{\Lambda}(\mathbf{D}^{1/2}\mathbf{V})^T.$$

Using the abbreviations

$$\mathbf{D}^{-1/2}\mathbf{V} = \boldsymbol{\Phi} = (\boldsymbol{\varphi}_1, \dots, \boldsymbol{\varphi}_n) \text{ and } \mathbf{D}^{1/2}\mathbf{V} = \boldsymbol{\Psi} = (\boldsymbol{\psi}_1, \dots, \boldsymbol{\psi}_n)$$

we get

$$\mathbf{M} = \boldsymbol{\Phi}\boldsymbol{\Lambda}\boldsymbol{\Psi}^T. \tag{4.17}$$

Matrices $\boldsymbol{\Phi}$ and $\boldsymbol{\Psi}$ are orthogonal in the sense

$$\boldsymbol{\Psi}^T\boldsymbol{\Phi} = \mathbf{I}_n, \text{ or equivalently } \boldsymbol{\psi}_i^T\boldsymbol{\varphi}_j = \delta_{ij}, \; i, j = 1, \dots, n. \tag{4.18}$$

Furthermore, λ_k are the eigenvalues of $\mathbf{M}$ with respect to right and left eigenvectors $\boldsymbol{\varphi}_k$ and $\boldsymbol{\psi}_k$, respectively, i.e.,

$$\mathbf{M}\boldsymbol{\varphi}_k = \lambda_k\boldsymbol{\varphi}_k \text{ and } \boldsymbol{\psi}_k^T\mathbf{M} = \lambda_k\boldsymbol{\psi}_k^T. \tag{4.19}$$

In summary,

$$\mathbf{M} = \sum_{k=1}^{n} \lambda_k \boldsymbol{\varphi}_k \boldsymbol{\psi}_k^T$$

so that by (4.18) we infer

$$\mathbf{M}^t = \sum_{k=1}^{n} \lambda_k^t \boldsymbol{\varphi}_k \boldsymbol{\psi}_k^T$$

and

$$\mathbf{u}_i = \mathrm{e}_i^T\,\mathbf{M}^t = \sum_{k=1}^{n} \lambda_k^t\, \mathrm{e}_i^T\, \boldsymbol{\varphi}_k \boldsymbol{\psi}_k^T = \sum_{k=1}^{n} \lambda_k^t \varphi_{ki} \boldsymbol{\psi}_k^T \tag{4.20}$$

with the notation $\boldsymbol{\varphi}_k = (\varphi_{k1}, \dots, \varphi_{kn})$. Equation (4.20) gives a representation of $\mathrm{e}_i^T\,\mathbf{M}^t$ in terms of the orthogonal basis $\boldsymbol{\psi}_1, \dots, \boldsymbol{\psi}_n$ with coefficients $\lambda_k^t\varphi_{k,i}$, $k = 1, \dots, n$. The coefficients define the so called *diffusion map*.

Recall that Lemma 2.22 sheds light on the eigenvalues and eigenvectors of the transition matrix $\mathbf{M}$. It states that all eigenvalues $\lambda_1, \dots, \lambda_n$ of $\mathbf{M}$ satisfy $|\lambda_k| \le 1$. Furthermore, $\mathbf{M}\mathbf{1}_n = \mathbf{1}_n$ such that 1 is an eigenvalue of $\mathbf{M}$.

Some properties of the transition matrix $\mathbf{M}$ carry over to the normalized version $\mathbf{S}$ and its spectral decomposition.

Lemma 4.7 *Matrices* $\mathbf{S}$ *in (4.16) and* $\mathbf{M}$ *in (4.15) have the same eigenvalues. Without loss of generality, the first column* $\boldsymbol{\varphi}_1$ *of* $\boldsymbol{\Phi} = \mathbf{D}^{-1/2}\mathbf{V}$ *is the all-one vector* $\mathbf{1}_n$.

Proof Assume that λ is an eigenvalue of $\mathbf{S}$ with eigenvector $\mathbf{x}$ so that

$$\mathbf{S}\mathbf{x} = \mathbf{D}^{1/2}\mathbf{M}\,\mathbf{D}^{-1/2}\mathbf{x} = \lambda\mathbf{x}.$$

Hence

$$\mathbf{M}\mathbf{D}^{-1/2}\mathbf{x} = \lambda \mathbf{D}^{-1/2}\mathbf{x},$$

which entails that λ is an eigenvalue of $\mathbf{M}$ with eigenvector $\mathbf{D}^{-1/2}\mathbf{x}$. Vice versa, if λ is an eigenvalue of $\mathbf{M}$ with eigenvector $\mathbf{y}$, write $\mathbf{y} = \mathbf{D}^{-1/2}\mathbf{x}$ to see that λ is an eigenvalue of $\mathbf{S}$ with eigenvector $\mathbf{x}$.

Now let $\lambda = 1$ be the largest eigenvalue of $\mathbf{S}$, corresponding to eigenvector $\mathbf{v}_1$, such that

$$\mathbf{D}^{1/2}\mathbf{M}\mathbf{D}^{-1/2}\mathbf{v}_1 = \mathbf{v}_1 .$$

Then

$$\mathbf{M}\mathbf{D}^{-1/2}\mathbf{v}_1 = \mathbf{D}^{-1/2}\mathbf{v}_1 \text{ with } \mathbf{D}^{-1/2}\mathbf{v}_1 = \mathbf{1}_n .$$

Hence, the first column $\boldsymbol{\varphi}_1 = \mathbf{D}^{-1/2}\mathbf{v}_1$ of $\boldsymbol{\Phi} = (\boldsymbol{\varphi}_1, \ldots, \boldsymbol{\varphi}_n)$ is the normalized all-one vector $\frac{1}{\sqrt{n}}\mathbf{1}$, which completes the proof. ■

By Lemma 4.7 the eigenvalue $\lambda_1 = 1$ does not contain any information about the distance structure of the vertices. It appears with the constant vector $\boldsymbol{\varphi}_1 = \mathbf{1}_n$ in any instance and can hence be omitted from the diffusion map.

Definition 4.8 The diffusion map at time t is defined as

$$\boldsymbol{\Theta}_t(v_i) = \begin{pmatrix} \lambda_2^t \varphi_{2i} \\ \vdots \\ \lambda_n^t \varphi_{ni} \end{pmatrix}, \quad i = 1, \ldots, n.$$

Recall that for all eigenvalues we have $|\lambda_k| \le 1$. Thus, if $|\lambda_k|$ is small then λ_k^t decreases exponentially in magnitude such that very small values are achieved even for moderate $t \in \mathbb{N}$. This motivates truncating the diffusion map to $d < n - 1$ dimensions.

Definition 4.9 The diffusion map at time t truncated to $d < n - 1$ dimensions is defined as

$$\boldsymbol{\Theta}_t^{(d)}(v_i) = \begin{pmatrix} \lambda_2^t \varphi_{2i} \\ \vdots \\ \lambda_{d+1}^t \varphi_{d+1,i} \end{pmatrix}, \quad i = 1, \ldots, n.$$

The points $\boldsymbol{\Theta}_t^{(d)}(v_i)$ are an approximate embedding of $v_1, \ldots, v_n$ in a d-dimensional Euclidean space.

The connection between the Euclidean distance in the diffusion map coordinates (the diffusion distance) and the dissimilarity between the corresponding probability distributions is specified in the following.

Theorem 4.10 *For any pair of nodes v_i and v_j of the weighted graph it holds that*

$$\left\|\boldsymbol{\Theta}_t(v_i) - \boldsymbol{\Theta}_t(v_j)\right\|^2 = \sum_{\ell=1}^{n} \frac{1}{\deg(v_\ell)} \Big(P(X_t = v_\ell \mid X_0 = v_i) - P(X_t = v_\ell \mid X_0 = v_j)\Big)^2.$$

A compact proof of Theorem 4.10 can be found in [5].

A flexible method to construct a weighted graph representing some cloud of points $\mathbf{x}_1, \ldots, \mathbf{x}_n$ is by use of *kernels* $K_\varepsilon : \mathbb{R}^p \times \mathbb{R}^p \to [0, 1]$ with parameter ε. The vertices of the graph are associated to the data points. The weights w_{ij} of the corresponding edges are then defined by

$$w_{ij} = K_\varepsilon(\boldsymbol{x}_i, \boldsymbol{x}_j). \tag{4.21}$$

In order to get closer to the geodesic distances on the hidden manifold, an appropriate kernel must satisfy the following properties:

(i) symmetry $K_\varepsilon(\mathbf{x}, \mathbf{y}) = K_\varepsilon(\mathbf{y}, \mathbf{x})$ for all $\mathbf{x}, \mathbf{y} \in \mathbb{R}^p$,
(ii) nonnegativity $K_\varepsilon(\mathbf{x}, \mathbf{y}) \geq 0$ for all $\mathbf{x}, \mathbf{y} \in \mathbb{R}^p$, and
(iii) locality, i.e., $\|\mathbf{x} - \mathbf{y}\| \ll \varepsilon \Rightarrow K_\varepsilon(\mathbf{x}, \mathbf{y}) \to 1$ and $\|\mathbf{x} - \mathbf{y}\| \gg \varepsilon \Rightarrow K_\varepsilon(\mathbf{x}, \mathbf{y}) \to 0$ for all $\mathbf{x}, \mathbf{y} \in \mathbb{R}^p$ and $\varepsilon > 0$.

A common choice for a kernel is the so-called *Gaussian kernel* defined by

$$K_\varepsilon(\mathbf{x}, \mathbf{y}) = \exp\left(-\frac{\|\mathbf{x} - \mathbf{y}\|^2}{2\varepsilon^2}\right).$$

In this case the weights w_{ij} in (4.21) will be less than 0.011 if $\|\mathbf{x}_i - \mathbf{x}_j\| > 3\varepsilon$.

This method works particularly well if the structure of the point cloud is far from being linear (think of the Swiss roll example in Fig. 4.4).

4.4 Exercises

Exercise 4.1 Four observations in $\mathbb{R}^3$ are given as follows:

$$\mathbf{x}_1 = \begin{pmatrix} 1 \\ 2 \\ -3 \end{pmatrix}, \quad \mathbf{x}_2 = \begin{pmatrix} 3 \\ -1 \\ -2 \end{pmatrix}, \quad \mathbf{x}_3 = \begin{pmatrix} -4 \\ 2 \\ 2 \end{pmatrix}, \quad \mathbf{x}_4 = \begin{pmatrix} -3 \\ -1 \\ 4 \end{pmatrix}.$$

(a) Compute the sample mean $\bar{\mathbf{x}}$ and the sample covariance matrix $\mathbf{S}_4$.
(b) Apply PCA to determine the optimal orthogonal projection matrix $\mathbf{Q}$ for presenting the data in two dimensions.

Exercise 4.2 The sample covariance matrix $\mathbf{S}_n$ for n observations is given by

$$\mathbf{S}_n = \begin{pmatrix} 14 & -14 \\ -14 & 110 \end{pmatrix}.$$

(a) Calculate the spectral decomposition $\mathbf{S}_n = \mathbf{V}\,\boldsymbol{\Lambda}\,\mathbf{V}^T$ where $\boldsymbol{\Lambda}$ is the diagonal matrix of eigenvalues and $\mathbf{V}$ comprises the orthogonal eigenvectors.
(b) Apply PCA and determine the optimal projection matrix $\mathbf{Q}$ for representing the data on the real line.
(c) Determine the percentage of variance which is still preserved by the projected points $\mathbf{Q}\mathbf{x}_i$ by computing the ratio (4.4).

Exercise 4.3 Fix $p = 500$ as the dimension of the space $\mathbb{R}^p$. Suppose that signal data is generated in two one-dimensional spaces, modeled by $\sqrt{0.2}G_1\mathbf{v}_1$ and $\sqrt{0.5}G_2\mathbf{v}_2$, where $\mathbf{v}_1$ and $\mathbf{v}_2$ are orthogonal unit norm vectors in $\mathbb{R}^p$ and G_1, G_2 are independent standard normal random variables. Independent high-dimensional noise $\boldsymbol{U} \in \mathbb{R}^p$ is modeled as a standard normal random vector. Hence, the corresponding stochastic model is $\boldsymbol{X} = \boldsymbol{U} + \sqrt{0.2}G_1\mathbf{v}_1 + \sqrt{0.5}G_2\mathbf{v}_2$ with jointly independent $\boldsymbol{U}, G_1, G_2$, all having zero mean. The covariance matrix of $\boldsymbol{X}$ is known to be

$$\operatorname{Cov}(\boldsymbol{X}) = \mathbf{I}_p + 0.2\mathbf{v}_1\mathbf{v}_1^T + 0.5\mathbf{v}_2\mathbf{v}_2^T.$$

Suppose that $\boldsymbol{X}_1, \ldots, \boldsymbol{X}_n$ are i.i.d. random vectors distributed as $\boldsymbol{X}$.

(a) Determine the minimum number n of samples such that approximately only the dominant eigenvalue is sticking out of the spectrum of the empirical covariance matrix S_n. Calculate the inner product $\langle \mathbf{v}_2, \mathbf{v}_{\text{dom}} \rangle$ for this case.
(b) Determine the minimum number n of samples such that both dominant eigenvalues are sticking out. Calculate the inner products $\langle \mathbf{v}_2, \mathbf{v}_{\text{dom}} \rangle$ for this case. Sketch the Marčenko-Pastur density for the latter case along with both dominant eigenvalues of the sample covariance matrix $\mathbf{S}_n$.

Exercise 4.4 Let $G = (V, E, \mathbf{W})$ be a weighted graph with a symmetric weight matrix $\mathbf{W}$. The transition matrix $\mathbf{M}$ of a random walk on this graph is defined in (4.15).

(a) Prove that $\mathbf{M}$ has multiple eigenvalues equal to one if and only if the graph is disconnected.
(b) If the underlying graph G is connected, prove that $\mathbf{M}$ has an eigenvalue equal to -1 if and only if the graph is bipartite.

Exercise 4.5 Given a data matrix $\mathbf{X} = (\mathbf{x}_1, \ldots, \mathbf{x}_n) \in \mathbb{R}^{p\times n}$, which is centered, i.e., $\mathbf{X}\mathbf{E}_n = \mathbf{X}$ or equivalently $\bar{\mathbf{x}} = \frac{1}{n}\sum_{i=1}^n \mathbf{x}_i = \mathbf{0}_p$.

(a) Show that $\mathbf{X}\mathbf{X}^T$ and $\mathbf{X}^T\mathbf{X}$ have the same eigenvalues up to the multiplicity of eigenvalue 0.
(b) For an embedding in k dimensions, $k < p$, PCA and MDS are applied. Show that up to an orthogonal transformation both methods yield the same configuration of n points in $\mathbb{R}^k$.

Exercise 4.6 Consider the set of dissimilarity matrices $\mathcal{U}_n$ in (4.10), also called *hollow* matrices.

(a) Show that $\mathcal{U}_n$ is a vector space.
(b) Show that $\|\mathbf{E}_n\mathbf{A}\mathbf{E}_n\|_F$, $\mathbf{A} \in \mathcal{U}_n$, defines a norm on $\mathcal{U}_n$.

Exercise 4.7 Assume n equispaced objects at equal distance 1 such that the dissimilarity $\mathbf{\Delta}$ is

$$\mathbf{\Delta} = \mathbf{1}_{n\times n} - \mathbf{I}_n = \begin{pmatrix} 0 & 1 & \cdots & 1 \\ 1 & 0 & \cdots & 1 \\ \vdots & & \ddots & \vdots \\ 1 & 1 & \cdots & 0 \end{pmatrix}.$$

Show that there is a Euclidean embedding in $n-1$ dimensions but no embedding in fewer dimensions less than $n-1$.

Exercise 4.8 In this exercise, we examine what happens if the data set is not large enough to detect its geometry. Consider the following five points in $\mathbb{R}^2$, see Fig. 4.5.

$$\mathbf{a} = \begin{pmatrix} 0 \\ 3 \end{pmatrix}, \; \mathbf{b} = \begin{pmatrix} 2 \\ 0 \end{pmatrix}, \; \mathbf{c} = \begin{pmatrix} -3 \\ 0 \end{pmatrix}, \; \mathbf{d} = \begin{pmatrix} 0 \\ -1 \end{pmatrix}, \; \mathbf{e} = \begin{pmatrix} -4 \\ -4 \end{pmatrix}.$$

(a) In order to find a one-dimensional embedding, construct the weighted graph for the IsoMap algorithm by using the 1-nearest neighbor criterion, the 2-nearest neighbors criterion and the mutual distances as weights. Estimate the geodesic distance of **e** and **d**.
(b) Construct the weighted graph for the IsoMap algorithm using ε as cutting bound for the mutual distances. Discuss the choice of ε.

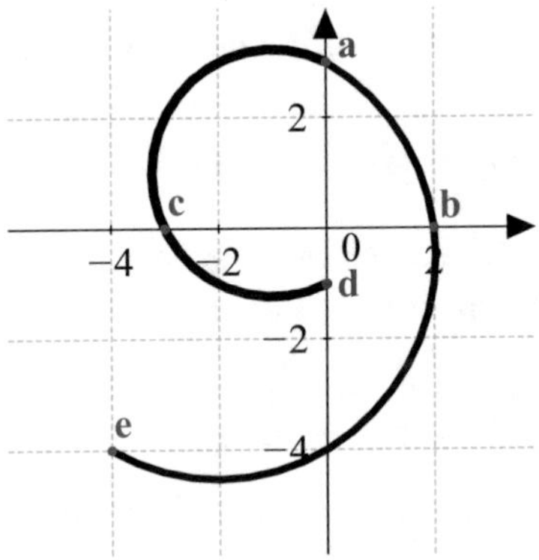

Fig. 4.5 Four points embedded in a one-dimensional manifold

Exercise 4.9 Three data points in $\mathbb{R}^3$ are given as follows

$$\mathbf{x}_1 = (1, -1, 1)^T, \ \mathbf{x}_2 = (-1, -1, -1)^T, \ \mathbf{x}_3 = (-1, 1, -1)^T,$$

and a kernel function $K(\mathbf{x}_i, \mathbf{x}_j) = \max\{1 - \frac{1}{6}\|\mathbf{x}_j - \mathbf{x}_i\|_2^2, 0\}$.

(a) For the random walk of the diffusion map a weight matrix $\mathbf{W}$ is needed. With the information above, calculate the missing weights of the following weight matrix $\mathbf{W} \in \mathbb{R}^{3\times 3}$:

$$\mathbf{W} = \begin{pmatrix} 1 & w_{12} & 0 \\ w_{21} & 1 & w_{23} \\ 0 & w_{32} & 1 \end{pmatrix}$$

(b) Compute the eigenvalues and the left and right eigenvectors in (4.19).

Chapter 5
Classification and Clustering

Classifying objects according to certain features is one of the fundamental problems in machine learning. Binary classification by supervised learning will be the topic of Chap. 6. In this chapter we will start with some elementary classification rules which are derived by a training set. The goal is to find a classifier that predicts the class correspondence of future observations as accurately as possible.

5.1 Discriminant Analysis

The global setup is as follows. Given k *populations*, *groups* or *classes* $\mathcal{C}_1, \dots, \mathcal{C}_k$, we wish to determine a partitioning

$$\mathcal{R}_1, \dots, \mathcal{R}_k \subseteq \mathbb{R}^p, \quad \bigcup_{i=1}^{k} \mathcal{R}_i = \mathbb{R}^p, \quad \mathcal{R}_i \cap \mathcal{R}_j = \emptyset \, \forall i \neq j,$$

and a *discriminant rule* to decide that a new observation $\mathbf{x}$ belongs to $\mathcal{C}_i$ if $\mathbf{x} \in \mathcal{R}_i$.

5.1.1 Fisher's Linear Discriminant Rule

Given a training set of elements $\mathbf{x}_1, \dots, \mathbf{x}_n \in \mathbb{R}^p$ with known class affiliation and some further observation $\mathbf{x} \in \mathbb{R}^p$. The objective is to determine a linear discriminant rule $\mathbf{a}^T\mathbf{x}$ such that $\mathbf{x}$ can be allocated to some class optimally.

Figure 5.1 shows an example with three classes of points in two dimensions and a one-dimensional representation. Keep in mind that if $\|\mathbf{a}\| = 1$ then $\mathbf{a}\mathbf{a}^T$ is an orthogonal projection matrix onto the one-dimensional line spanned by vector $\mathbf{a}$,

R. Mathar et al., *Fundamentals of Data Analytics*,
https://doi.org/10.1007/978-3-030-56831-3_5

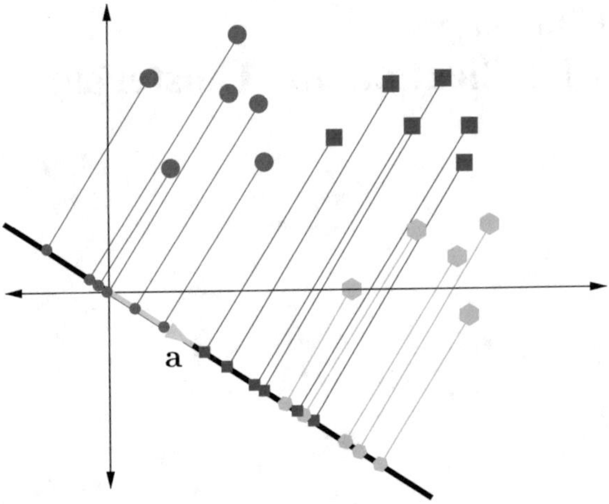

Fig. 5.1 Three classes of points in two dimensions and an optimal one-dimensional representation for discrimination. There is clearly an overlap between the projected blue and green data, although the classes are well separable in two dimensions. By dimensionality reduction one may always loose information

see Lemma 2.35. The projected point for some $\mathbf{x} \in \mathbb{R}^p$ is $\mathbf{a}\mathbf{a}^T\mathbf{x}$, which shows that $\mathbf{a}^T\mathbf{x}$ is the length of the projected point into the direction of $\mathbf{a}$.

We denote the projections by

$$y_i = \mathbf{a}^T\mathbf{x}_i,\ i = 1, \dots, n\,.$$

Intuitively, the optimum $\mathbf{a}$ is determined by maximizing the quotient of the sum of squares (SSQ) between groups over the sum of squares within groups. Making the nominator large and the denominator small yields a large value of the quotient. With the notation $\bar{y} = \frac{1}{n}\sum_{i=1}^{n} y_i$ and $\bar{y}_\ell = \frac{1}{n_\ell}\sum_{i\in C_\ell} y_i$ and n_ℓ the number of points in class ℓ this can be formalized as

$$\max_{\mathbf{a}\in\mathbb{R}} \left\{ \frac{\text{SSQ between groups}}{\text{SSQ within groups}} \right\} = \max_{\mathbf{a}\in\mathbb{R}} \left\{ \frac{\sum_{\ell=1}^{k} n_\ell(\bar{y}_\ell - \bar{y})^2}{\sum_{\ell=1}^{k}\sum_{j\in C_\ell}(y_j - \bar{y}_\ell)^2} \right\}.$$

In order to fully develop this optimization problem for a given training set $\mathbf{X} = [\mathbf{x}_1, \dots, \mathbf{x}_n] \in \mathbb{R}^{p\times n}$ of k nonempty groups $C_1, \dots, C_k$ we need the following notation. Firstly, n_ℓ denotes the number of elements in class C_ℓ and $\mathbf{X}_\ell = [\mathbf{x}_j]_{j\in C_\ell} \in \mathbb{R}^{p\times n_\ell}$ the submatrix combined of elements in group C_ℓ. The corresponding averages are

$$\bar{\mathbf{x}} = \frac{1}{n}\sum_{i=1}^{n} \mathbf{x}_i \quad \text{(total average)}$$

and

$$\bar{\mathbf{x}}_\ell = \frac{1}{n_\ell}\sum_{j\in C_\ell} \mathbf{x}_j \quad \text{(average over class } \ell\text{)}\,.$$

Let $\mathbf{a} \in \mathbb{R}^p$ be the optimization variable and

$$\mathbf{y} = (y_1, \ldots, y_n) = \mathbf{a}^T\mathbf{X} \in \mathbb{R}^n$$

the n-dimensional row-vector of transformed values $\mathbf{a}^T\mathbf{x}_i$. With $\mathbf{y}_\ell = (y_j)_{j\in C_\ell}$ and

$$\bar{y} = \frac{1}{n}\sum_{i=1}^{n} y_i \quad \text{and} \quad \bar{y}_\ell = \frac{1}{n_\ell}\sum_{j\in C_\ell} y_j$$

we first compute the sum of squares. First note that the mixed terms in the outer sum below are zero:

$$\begin{aligned}\sum_{i=1}^{n}(y_i - \bar{y})^2 &= \sum_{\ell=1}^{k}\sum_{j\in C_\ell}(y_j - \bar{y}_\ell + \bar{y}_\ell - \bar{y})^2 \\ &= \sum_{\ell=1}^{k}\Big(\sum_{j\in C_\ell}(y_j - \bar{y}_\ell)^2 + \sum_{j\in C_\ell}(\bar{y}_\ell - \bar{y})^2\Big) \\ &= \sum_{\ell=1}^{k}\sum_{j\in C_\ell}(y_j - \bar{y}_\ell)^2 \;+\; \sum_{\ell=1}^{k} n_\ell(\bar{y}_\ell - \bar{y})^2 \,. \end{aligned} \tag{5.1}$$

The first addend is the sum of squares within groups, the second one is the sum of squares between groups, both corresponding to the amount of variation.

In matrix notation we get an equivalent representation, amenable to the optimization problems in Chap. 2. Recall from (4.12) that

$$\mathbf{E}_n = \mathbf{I}_n - \tfrac{1}{n}\mathbf{1}_n\mathbf{1}_n^T = \mathbf{I}_n - \tfrac{1}{n}\mathbf{1}_{n\times n}\,.$$

For the first term in (5.1) we can write

$$\begin{aligned}\sum_{\ell=1}^{k}\sum_{j\in C_\ell}(y_j - \bar{y}_\ell)^2 &= \sum_{\ell=1}^{k}\mathbf{y}_\ell^T\mathbf{E}_{n_\ell}\mathbf{y}_\ell = \sum_{\ell=1}^{k}\mathbf{a}^T\mathbf{X}_\ell\mathbf{E}_{n_\ell}\mathbf{X}_\ell^T\mathbf{a} \\ &= \mathbf{a}^T\Big[\sum_{\ell=1}^{k}\mathbf{X}_\ell\mathbf{E}_{n_\ell}\mathbf{X}_\ell^T\Big]\mathbf{a} = \mathbf{a}^T\mathbf{W}\mathbf{a},\end{aligned} \tag{5.2}$$

say, where matrix $\mathbf{W} = \sum_{\ell=1}^{k}\mathbf{X}_\ell\mathbf{E}_{n_\ell}\mathbf{X}_\ell^T \in \mathbb{R}^{p\times p}$ stands for within classes SSQ.

For the second term in (5.1) we get

$$\sum_{\ell=1}^{k} n_\ell(\bar{y}_\ell - \bar{y})^2 = \sum_{\ell=1}^{k} n_\ell\big(\mathbf{a}^T(\bar{\mathbf{x}}_\ell - \bar{\mathbf{x}})\big)^2 = \sum_{\ell=1}^{k} n_\ell\, \mathbf{a}^T(\bar{\mathbf{x}}_\ell - \bar{x})(\bar{\mathbf{x}}_\ell - \bar{\mathbf{x}})^T\mathbf{a}$$
$$= \mathbf{a}^T\Big[\sum_{\ell=1}^{k} n_\ell\,(\bar{\mathbf{x}}_\ell - \bar{\mathbf{x}})(\bar{\mathbf{x}}_\ell - \bar{\mathbf{x}})^T\Big]\mathbf{a} = \mathbf{a}^T\mathbf{B}\mathbf{a}, \tag{5.3}$$

say, where matrix $\mathbf{B} = \sum_{\ell=1}^{k} n_\ell\,(\bar{\mathbf{x}}_\ell - \bar{\mathbf{x}})(\bar{\mathbf{x}}_\ell - \bar{\mathbf{x}})^T$ stands for between classes SSQ.

Linear discriminant analysis requires to maximize the ratio

$$\max_{\mathbf{a}\in\mathbb{R}^p} \frac{\mathbf{a}^T\mathbf{B}\mathbf{a}}{\mathbf{a}^T\mathbf{W}\mathbf{a}}. \tag{5.4}$$

Theorem 5.1 *If* $\mathbf{W}$ *is invertible, the maximum of (5.4) is attained at the eigenvector* $\mathbf{a}^*$ *of* $\mathbf{W}^{-1}\mathbf{B}$ *corresponding to the largest eigenvalue.*

Proof First note that both $\mathbf{B}$ and $\mathbf{W}$ are symmetric. Further, $\mathbf{W}^{-1/2}\mathbf{B}\mathbf{W}^{-1/2}$ and $\mathbf{W}^{-1}\mathbf{B}$ have the same eigenvalues since $\mathbf{W}^{-1/2}\mathbf{B}\mathbf{W}^{-1/2}\mathbf{v} = \lambda\mathbf{v}$ if and only if $\mathbf{W}^{-1}\mathbf{B}\mathbf{W}^{-1/2}\mathbf{v} = \lambda\mathbf{W}^{-1/2}\mathbf{v}$.

By Theorem 2.26, particularly (2.17) it follows that

$$\max_{\mathbf{a}\in\mathbb{R}^p} \frac{\mathbf{a}^T\mathbf{B}\mathbf{a}}{\mathbf{a}^T\mathbf{W}\mathbf{a}} = \max_{\mathbf{b}\in\mathbb{R}^p} \frac{\mathbf{b}^T\mathbf{W}^{-1/2}\mathbf{B}\mathbf{W}^{-1/2}\mathbf{b}}{\mathbf{b}^T\mathbf{b}} = \lambda_{\max}\big(\mathbf{W}^{-1/2}\mathbf{B}\mathbf{W}^{-1/2}\big),$$

which completes the proof. ■

The application of discriminant rule (5.4) starts from a training set $\mathbf{x}_1, \ldots, \mathbf{x}_n$ with known class affiliation. First, the optimal $\mathbf{a}$ is computed according to Theorem 5.1. A new observation $\mathbf{x}$ with unknown class affiliation is allocated to the class $\mathcal{C}_\ell$ if

$$|\mathbf{a}^T\mathbf{x} - \mathbf{a}^T\bar{\mathbf{x}}_\ell| \le |\mathbf{a}^T\mathbf{x} - \mathbf{a}^T\bar{\mathbf{x}}_j| \text{ for all } j = 1, \ldots, k.$$

The special case of $k = 2$ classes $\mathcal{C}_1$ and $\mathcal{C}_2$ can explicitly be derived from the general approach. We assume that there are two groups of sizes n_1 and n_2 with $n = n_1 + n_2$. Since $n_1\bar{\mathbf{x}}_1 + n_2\bar{\mathbf{x}}_2 = (n_1 + n_2)\bar{\mathbf{x}}$ we obtain after some tedious but elementary algebra

$$\begin{aligned}\mathbf{B} &= n_1(\bar{\mathbf{x}}_1 - \bar{\mathbf{x}})(\bar{\mathbf{x}}_1 - \bar{\mathbf{x}})^T + n_2(\bar{\mathbf{x}}_2 - \bar{\mathbf{x}})(\bar{\mathbf{x}}_2 - \bar{\mathbf{x}})^T \\ &= \frac{n_1 n_2}{n}(\bar{\mathbf{x}}_1 - \bar{\mathbf{x}}_2)(\bar{\mathbf{x}}_1 - \bar{\mathbf{x}}_2)^T \\ &= \frac{n_1 n_2}{n}\,\mathbf{d}\mathbf{d}^T, \text{ say.}\end{aligned}$$

Matrix $\mathbf{B}$ has rank 1, i.e., at most one eigenvalue is different from zero. The positive eigenvalue of $\mathbf{W}^{-1}\mathbf{B}$ is

$$\lambda_1 = \mathrm{tr}\left(\mathbf{W}^{-1}\mathbf{B}\right) = \frac{n_1 n_2}{n}\,\mathbf{d}^T\mathbf{W}^{-1}\mathbf{d}$$

with eigenvector $\mathbf{a} = \mathbf{W}^{-1}\mathbf{d}$. This holds since

$$\mathbf{W}^{-1}\mathbf{B}\,\mathbf{W}^{-1}\mathbf{d} = \frac{n_1 n_2}{n}\,\mathbf{W}^{-1}\mathbf{d}\mathbf{d}^T\,\mathbf{W}^{-1}\mathbf{d} = \lambda_1\mathbf{W}^{-1}\mathbf{d}.$$

The discriminant rule for $k = 2$ boils down to allocate a new observation $\mathbf{x}$ to class $\mathcal{C}_1$ if

$$\mathbf{d}^T\mathbf{W}^{-1}\left(\mathbf{x} - \tfrac{1}{2}(\bar{\mathbf{x}}_1 + \bar{\mathbf{x}}_2)\right) > 0 \tag{5.5}$$

and otherwise to class $\mathcal{C}_2$.

Note that the optimal $\mathbf{a} = \mathbf{W}^{-1}(\bar{\mathbf{x}}_1 - \mathbf{x}_2)$ is normal to a discrimination hyperplane of dimension $p - 1$ between the classes.

Fishers approach is distribution free, i.e., no knowledge about the distribution of the training set is needed. Instead, the rule relies on the general principle that the between-groups SSQ should be large relative to the within-groups SSQ.

5.1.2 Gaussian ML Discriminant Rule

In contrast to the previous section we now assume that the distribution of points in each class is known. Normally this assumption is parametric and the parameters of the corresponding distributions are estimated from the training set.

A typical example is to suppose that the distribution of each class $\mathcal{C}_\ell$ is Gaussian with expectation vector $\boldsymbol{\mu}_\ell$ and covariance matrix $\boldsymbol{\Sigma}_\ell$, $\ell = 1, \dots, k$. The corresponding densities are

$$f_\ell(\mathbf{u}) = \frac{1}{(2\pi)^{p/2}|\boldsymbol{\Sigma}_\ell|^{1/2}}\exp\left\{-\frac{1}{2}\left(\mathbf{u} - \boldsymbol{\mu}_\ell\right)^T\boldsymbol{\Sigma}_\ell^{-1}\left(\mathbf{u} - \boldsymbol{\mu}_\ell\right)\right\}. \tag{5.6}$$

The maximum likelihood rule (ML rule) allocates a new observation $\mathbf{x}$ to the class $\mathcal{C}_\ell$ which maximizes the likelihood or log-likelihood function $L_\ell(\mathbf{x}) = \max_j L_j(\mathbf{x})$. Hence, given some $\mathbf{x}$, our objective is to maximize $f_\ell(\mathbf{x})$ from (5.6) over $\ell = 1, \dots, k$.

Theorem 5.2 *Assume that the class distributions are normal with densities (5.6) and* $\mathbf{x}$ *is an observation to be classified.*

(a) *If* $\boldsymbol{\Sigma}_\ell = \boldsymbol{\Sigma}$ *for all* ℓ, *then the ML rule allocates* $\mathbf{x}$ *to the class* $\mathcal{C}_\ell$ *which attains the minimum over the squared* Mahalanobis distances $\left(\mathbf{x} - \boldsymbol{\mu}_\ell\right)^T\boldsymbol{\Sigma}^{-1}\left(\mathbf{x} - \boldsymbol{\mu}_\ell\right)$, $\ell = 1 \dots, k$.

(b) *If there are* $k = 2$ *classes and* $\boldsymbol{\Sigma}_1 = \boldsymbol{\Sigma}_2 = \boldsymbol{\Sigma}$, *then the ML rule allocates* $\mathbf{x}$ *to class* $\mathcal{C}_1$ *if* $\boldsymbol{\alpha}^T(\mathbf{x} - \boldsymbol{\mu}) > 0$ *and otherwise to* $\mathcal{C}_2$, *where* $\boldsymbol{\alpha} = \boldsymbol{\Sigma}^{-1}(\boldsymbol{\mu}_1 - \boldsymbol{\mu}_2)$ *and* $\boldsymbol{\mu} = \frac{1}{2}(\boldsymbol{\mu}_1 + \boldsymbol{\mu}_2)$.

Proof Part (a) is obvious. To demonstrate part (b) we observe that in formula (5.6) $f_1(\mathbf{x}) > f_2(\mathbf{x})$ if and only if

$$(\mathbf{x} - \boldsymbol{\mu}_1)^T \boldsymbol{\Sigma}^{-1} (\mathbf{x} - \boldsymbol{\mu}_1) < (\mathbf{x} - \boldsymbol{\mu}_2)^T \boldsymbol{\Sigma}^{-1} (\mathbf{x} - \boldsymbol{\mu}_2),$$

which after some algebra turns out to be equivalent to

$$(\boldsymbol{\mu}_1 - \boldsymbol{\mu}_2)^T \boldsymbol{\Sigma}^{-1} \big(\mathbf{x} - \tfrac{1}{2}(\boldsymbol{\mu}_1 + \boldsymbol{\mu}_2)\big) > 0$$

and completes the proof. ■

The rule in part (b) above is analogous to Fisher's discriminant rule (5.5) with $\boldsymbol{\mu}_1, \boldsymbol{\mu}_2$ and $\boldsymbol{\Sigma}$ substituting the estimators $\bar{\mathbf{x}}_1, \bar{\mathbf{x}}_2$ and $\mathbf{W}$, respectively.

This brings us to the point how to apply the distribution based rule in practice, since the parameters $\boldsymbol{\mu}_\ell$ and $\boldsymbol{\Sigma}_\ell$ are normally unknown. They are first estimated by the training set with known class allocation and then substituted by the estimated values $\hat{\boldsymbol{\mu}}_\ell$ and $\hat{\boldsymbol{\Sigma}}_\ell$.

5.2 Cluster Analysis

Clustering intends to find meaningful groups within the data such that similar objects will be members of the same group and can be represented by a single cluster representative. This rather intuitive idea will be made mathematically precise by specifying different concepts of similarity and metrics of success. Clustering is an unsupervised learning method. The aim is to split the data into k groups which represent certain common features of the members in a meaningful way.

5.2.1 *k-Means Clustering*

As usual we start from p-dimensional data $\mathbf{x}_1, \ldots, \mathbf{x}_n \in \mathbb{R}^p$. Assuming that there are k groups so that points within the groups are near to each other, our aim is to find a partitioning of subsets $\mathcal{S}_1, \ldots, \mathcal{S}_k$ of $\{1, \ldots, n\}$ and centers $\boldsymbol{\mu}_1, \ldots, \boldsymbol{\mu}_k \in \mathbb{R}^p$ as a solution of

$$\min_{\substack{\text{partitioning } \mathcal{S}_1,\ldots,\mathcal{S}_k \\ \boldsymbol{\mu}_1,\ldots,\boldsymbol{\mu}_k \in \mathbb{R}^p}} \sum_{\ell=1}^{k} \sum_{i \in \mathcal{S}_\ell} \left\| \mathbf{x}_i - \boldsymbol{\mu}_\ell \right\|^2 .$$

This problem is NP-hard. Roughly speaking, finding a solution demands exponentially increasing computational effort.

However, given a partitioning, the optimal centers are the group averages

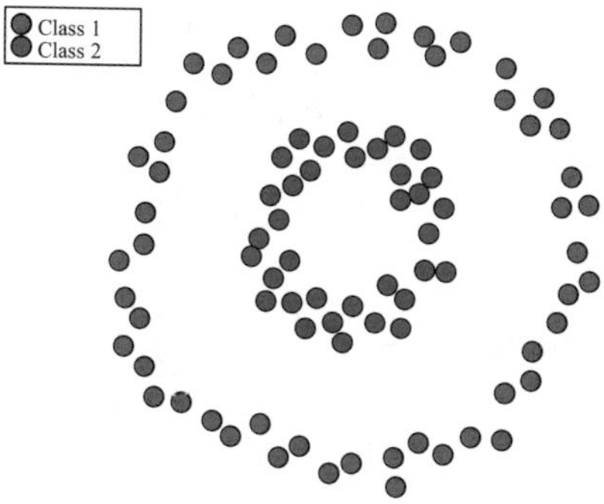

Fig. 5.2 Two data patterns which would not be classifiable by Lloyd's algorithm. Human observers easily detect two rings of points, a meaningful clustering for the data w.r.t. nearness

$$\boldsymbol{\mu}_\ell = \bar{\mathbf{x}}_\ell = \frac{1}{n_\ell} \sum_{i \in \mathcal{S}_\ell} \mathbf{x}_i \,.$$

as follows from Theorem 3.5, where $n_\ell = |\mathcal{S}_\ell|$ is the number of elements in each cluster. This motivates the following greedy k-means algorithm, also called Lloyd's algorithm [22].
Alternate between the steps (i) and (ii) until convergence:

(i) Update the partitioning: Given the centers $\boldsymbol{\mu}_1, \dots, \boldsymbol{\mu}_k \in \mathbb{R}^p$. For all $i = 1, \dots, n$ assign $\mathbf{x}_i$ to cluster C_ℓ if its distance to $\boldsymbol{\mu}_\ell$ is minimal, i.e. $\ell = \arg\min_j \|\mathbf{x}_i - \boldsymbol{\mu}_j\|^2$ and $\mathcal{S}_\ell \cup \{i\} \to \mathcal{S}_\ell$.
(ii) Update the centers: Given the partitioning $\mathcal{S}_1, \dots, \mathcal{S}_k \subseteq \{1, \dots, n\}$, determine the centers $\boldsymbol{\mu}_\ell = \frac{1}{|\mathcal{S}_\ell|} \sum_{i \in \mathcal{S}_\ell} \mathbf{x}_i$ for all $\ell = 1, \dots, k$.

Remark 5.3 (a) Lloyd's algorithm needs to know the number of clusters a-priory, which is often unknown.
(b) A concept of measuring closeness is essential. Generalizations would only work for metric spaces.
(c) The algorithm may stop in a suboptimal solution. It is of greedy type, finding an optimal solution in each step. This does not necessarily lead to a global optimum.
(d) The algorithm finds clusters which are pairwise linearly separable. In many cases this is not arguable by the structure of the data, cf. Fig. 5.2. For this example any solution of Lloyd's algorithm will be meaningless.

5.2.2 *Spectral Clustering*

Diffusion maps can cluster nonlinearly separable patterns. The approach is analogous to Sect. 4.3.2. From given data $\mathbf{x}_1, \dots \mathbf{x}_n \in \mathbb{R}^p$ we construct a weighted graph $G =$

$(V, E, \mathbf{W})$. Each point $\mathbf{x}_i$ represents a vertex v_i. The edge weights $\mathbf{W} = (w_{ij})_{i,j}$ are given by

$$w_{ij} = K_\varepsilon(\mathbf{x}_i, \mathbf{x}_j)$$

with some kernel $K_\varepsilon : \mathbb{R}^p \times \mathbb{R}^p \to [0, 1]$, e.g., the Gauss-kernel

$$K_\varepsilon(\mathbf{x}_i, \mathbf{x}_j) = \exp\left(-\frac{\|\mathbf{x}_i - \mathbf{x}_j\|^2}{2\varepsilon^2}\right).$$

The distance $\|\mathbf{x}_i - \mathbf{x}_j\|$ herein can be substituted by any meaningful symmetric dissimilarity measure.

Consider a homogeneous random walk $X_t, t = 0, 1, 2, \ldots$, with transition matrix

$$\mathbf{M} = \big(m_{ij}\big)_{i,j=1,\ldots,n} = \mathbf{D}^{-1}\mathbf{W},$$

where $\mathbf{D} = \operatorname{diag}\big(\deg(v_1), \ldots, \deg(v_n)\big)$ denotes the diagonal matrix of node degrees, $\deg(v_i) = \sum_{j=1}^{n} w_{ij}$. It holds that

$$P\big(X_{t+1} = v_j \mid X_t = v_i\big) = \frac{w_{ij}}{\deg(v_i)} = m_{ij}\,.$$

Similarly to (4.17) we decompose

$$\mathbf{M} = \mathbf{\Phi}\mathbf{\Lambda}\mathbf{\Psi} = \sum_{\ell=1}^{n} \lambda_\ell \boldsymbol{\varphi}_\ell \boldsymbol{\psi}_\ell^T$$

with eigenvalue matrix $\mathbf{\Lambda} = \operatorname{diag}(\lambda_1, \ldots, \lambda_n)$, $\lambda_1 \geq \cdots \geq \lambda_n$, of $\mathbf{M}$ corresponding to biorthonormal matrices $\mathbf{\Phi} = (\boldsymbol{\varphi}_1, \ldots, \boldsymbol{\varphi}_n)$ and $\mathbf{\Psi} = (\boldsymbol{\psi}_1, \ldots, \boldsymbol{\psi}_n)$ composed of the right and left eigenvectors of $\mathbf{M}$, respectively. It follows for $t \in \mathbb{N}$ that

$$\mathbf{M}^t = \mathbf{\Phi}\mathbf{\Lambda}^t\mathbf{\Psi} = \sum_{\ell=1}^{n} \lambda_\ell^t \boldsymbol{\varphi}_\ell \boldsymbol{\psi}_\ell^T,$$

furthermore, for any $i \in \{1, \ldots, n\}$

$$\mathrm{e}_i^T\,\mathbf{M}^t = \sum_{\ell=1}^{n} \lambda_\ell^t\, \mathrm{e}_i^T\, \boldsymbol{\varphi}_\ell \boldsymbol{\psi}_\ell^T = \sum_{\ell=1}^{n} \lambda_\ell^t \varphi_{\ell i}\, \boldsymbol{\psi}_\ell^T \tag{5.7}$$

with the notation $\boldsymbol{\varphi}_\ell = (\varphi_{\ell 1}, \ldots, \varphi_{\ell n})$. Equation (5.7) is a representation of $\mathrm{e}_i^T\,\mathbf{M}^t$ in terms of the orthogonal basis $\boldsymbol{\psi}_1, \ldots, \boldsymbol{\psi}_n$ with coefficients $\lambda_\ell^t \varphi_{\ell,i}$, $\ell = 1, \ldots, n$. If vertex v_i and vertex v_j are close in terms of strong connectivity, then after a sufficient number of transition steps $\mathrm{e}_i^T\,\mathbf{M}^t$ and $\mathrm{e}_j^T\,\mathbf{M}^t$ will be similar.

The coefficients $\lambda_\ell^t \varphi_{\ell,i}$ define the *diffusion map*. Since the all-one vector is always an eigenvector of $\mathbf{M}$ corresponding to eigenvalue 1, no information is conveyed for $\ell = 1$, see Lemma 2.22. The diffusion map at time t truncated to d dimensions is, cf. Definition 4.9,

$$\Theta_t^{(d)}(v_i) = \begin{pmatrix} \lambda_2^t \varphi_{2i} \\ \vdots \\ \lambda_{d+1}^t \varphi_{d+1,i} \end{pmatrix}, \quad i = 1, \ldots, n.$$

We now aim at clustering the vertices of the graph into k clusters by means of the diffusion map after t steps. First, the $(k-1)$-dimensional diffusion map is computed

$$\Theta_t^{(k-1)}(v_i) = \begin{pmatrix} \lambda_2^t \varphi_{2i} \\ \vdots \\ \lambda_k^t \varphi_{k,i} \end{pmatrix}, \quad i = 1, \ldots, n,$$

to create a set of n points in $\mathbb{R}^{k-1}$. Subsequently, k-means clustering is used to find an appropriate clustering into k groups.

In particular, in the case of only $k = 2$ clusters with partitioning $\mathcal{S}$ and $\mathcal{S}^c$ we get

$$\Theta_t^{(1)}(v_i) \in \mathbb{R}, \quad i = 1, \ldots, n,$$

as numbers on the real line. A natural way of clustering is to define a threshold $q \in \mathbb{R}$ such that

$$i \in \mathcal{S} \text{ if } \Theta_t^{(1)}(v_i) \leq q$$

and $i \in \mathcal{S}^c$ otherwise.

5.2.3 *Hierarchical Clustering*

Another type of unsupervised clustering algorithms are the hierarchical clustering methods. Hierarchical clustering algorithms are mainly divided into agglomerative (bottom up) and divisive (top down) clustering. Agglomerative methods assign a cluster to each observation and then reduce the number of clusters by iteratively merging smaller clusters into larger ones. On the other hand, divisive methods start with one large cluster containing all the observations and iteratively divide it into smaller clusters.

Since the principles behind agglomerative and divisive clustering are quite analogous, in this section we only explain agglomerative algorithms. Agglomerative clustering is based on dissimilarities δ_{ij} between n objects $v_1, \ldots, v_n$ and corresponding symmetric dissimilarity matrix

$$\boldsymbol{\Delta} = \left(\delta_{ij}\right)_{i,j=1,\ldots,n}$$

with $\delta_{ij} = \delta_{ji} \geq 0$ and $\delta_{ii} = 0$ for all $i, j = 1, \ldots, n$.

We first define a *linkage function* between clusters $\mathcal{C}_k$ and $\mathcal{C}_\ell$.

$$d(\mathcal{C}_k, \mathcal{C}_\ell) = \begin{cases} \min\limits_{\substack{v_i \in \mathcal{C}_k \\ v_j \in \mathcal{C}_\ell}} \delta_{ij} & \text{(single linkage)} \\ \max\limits_{\substack{v_i \in \mathcal{C}_k \\ v_j \in \mathcal{C}_\ell}} \delta_{ij} & \text{(complete linkage)} \\ \frac{1}{|\mathcal{C}_k| \cdot |\mathcal{C}_\ell|} \sum\limits_{\substack{v_i \in \mathcal{C}_k \\ v_j \in \mathcal{C}_\ell}} \delta_{ij} & \text{(average linkage)} \end{cases}$$

The following algorithm successively clusters objects according to their similarity. The ones which are near will be combined at early stages of the algorithm. The user can decide about the level of putting objects into a joint cluster.

Algorithm *Agglomerative Clustering*

- Initialize clusters as singletons:
  For $i = 1$ to n assign $\mathcal{C}_i \leftarrow \{v_i\}$
- Initialize the set of clusters available for merging:
  $\mathcal{S} \leftarrow \{1, \ldots, n\}$
- repeat
  - pick two most similar clusters to merge:
    $(k, \ell) \leftarrow \arg\min_{k,\ell \in \mathcal{S}} d(\mathcal{C}_k, \mathcal{C}_\ell)$
  - create a new cluster:
    $\mathcal{C}_m \leftarrow \mathcal{C}_k \cup \mathcal{C}_\ell$
  - mark k and ℓ as unavailable:
    $\mathcal{S} \leftarrow \mathcal{S} \setminus \{k, \ell\}$
  - if $\mathcal{C}_m \neq \{v_1, \ldots v_n\}$ then mark m as available:
    $\mathcal{S} \leftarrow \mathcal{S} \cup \{m\}$
  - for each $i \in \mathcal{S}$ do:
    update the dissimilarity matrix $d(\mathcal{C}_i, \mathcal{C}_m)$

  until no more clusters are available for merging.

There are different ways of graphically representing the corresponding cluster tree. Figure 5.3 shows 5 objects which are successively clustered.

Figure 5.4 shows the clustering trees, the so-called *dendrograms* for a huge dissimilarity matrix on the recognition scheme for handwritten digits $0, \ldots, 9$. Complete linkage is applied, the classes are additionally discriminated by colors. Figure 5.5 shows the same data set when clustered by average linkage. Obviously complete linkage gives the better result of splitting into classes corresponding to 10 digits.

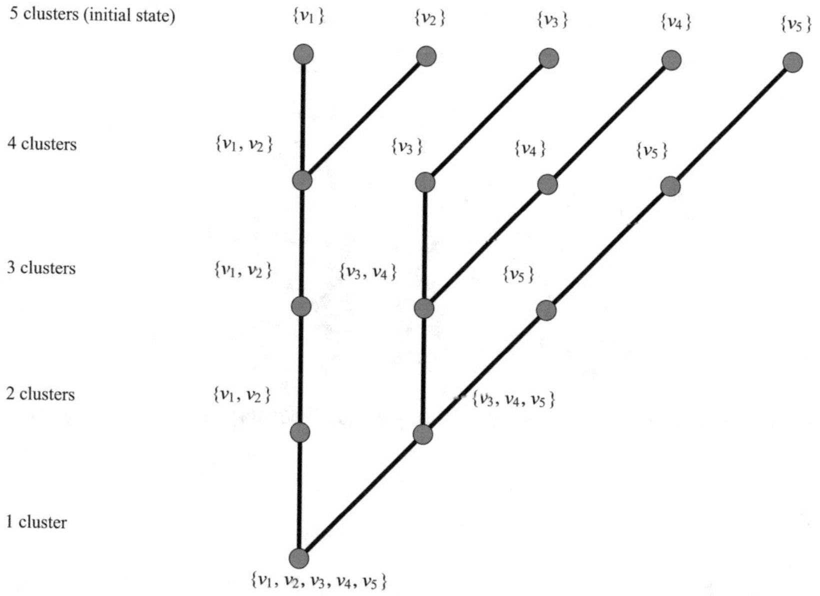

Fig. 5.3 Graphical example of agglomerative clustering with 5 objects ($n = 5$)

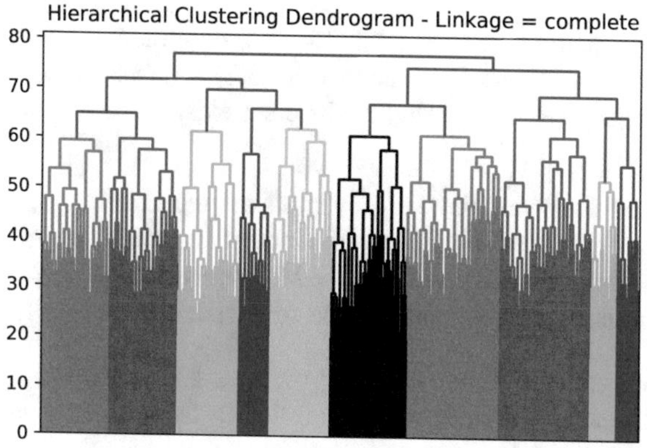

Fig. 5.4 Clustering of a data set on handwritten digits by the above algorithm using complete linkage

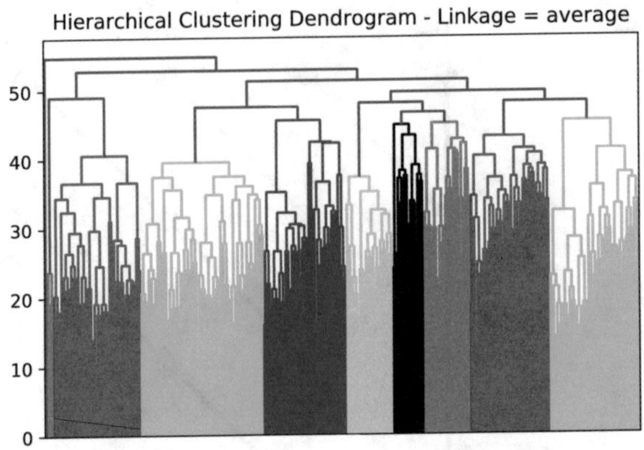

Fig. 5.5 Clustering of a data set on handwritten digits by the above algorithm using average linkage

5.3 Exercises

Exercise 5.1 Let $\mathbf{X} = [\mathbf{x}_1 \dots, \mathbf{x}_n]$ be a data matrix and $\bar{\mathbf{x}} = \frac{1}{n}\sum_{i=1}^n \mathbf{x}_i$. Show that

$$\mathbf{X}\,\mathbf{E}_n\mathbf{X}^T = \sum_{i=1}^n (\mathbf{x}_i - \bar{\mathbf{x}})(\mathbf{x}_i - \bar{\mathbf{x}})^T = \sum_{i=1}^n \mathbf{x}_i\mathbf{x}_i^T - n\bar{\mathbf{x}}\bar{\mathbf{x}}^T.$$

Exercise 5.2 Let $\mathbf{X} = [\mathbf{x}_1, \dots, \mathbf{x}_n] \in \mathbb{R}^{p\times n}$ be a data matrix. Within and between groups SSQ matrices $\mathbf{W}$ and $\mathbf{B}$ are defined in (5.2) and (5.3), respectively. We also define $\mathbf{S} = \mathbf{X}\,\mathbf{E}_n\mathbf{X}^T$. We assume that the inverses exist in the following.

(a) Show that $\mathbf{S} = \mathbf{B} + \mathbf{W}$.
(b) Show that the following three eigenvectors are the same:
- The eigenvector corresponding to the largest eigenvalue of $\mathbf{W}^{-1}\mathbf{B}$.
- The eigenvector corresponding to the largest eigenvalue of $\mathbf{W}^{-1}\mathbf{S}$.
- The eigenvector corresponding to the smallest eigenvalue of $\mathbf{S}^{-1}\mathbf{W}$.

Exercise 5.3 Consider training data $\mathbf{x}_1, \dots, \mathbf{x}_n \in \mathbb{R}^p$ generated by random vectors $X_1, \dots, X_n$ each with known class affiliation from a partitioning $\mathcal{S}_1, \dots, \mathcal{S}_k$ of $\{1, \dots, n\}$ as follows

$$X_i \sim \mathcal{N}_p\big(\boldsymbol{\mu}_\ell, \boldsymbol{\Sigma}\big), \text{ if } i \in \mathcal{S}_\ell.$$

with unknown expectation $\boldsymbol{\mu}_\ell$ and covariance matrix $\boldsymbol{\Sigma}$.

To apply the Gaussian ML discriminant rule one needs the ML estimator of $\boldsymbol{\mu}_\ell$ and $\boldsymbol{\Sigma}$. Show that

$$\widehat{\boldsymbol{\Sigma}} = \frac{1}{n} \sum_{\ell=1}^{k} \sum_{i \in \mathcal{S}_\ell} (\mathbf{x}_i - \bar{\mathbf{x}}_\ell)(\mathbf{x}_i - \bar{\mathbf{x}}_\ell)^T = \frac{1}{n} \mathbf{W}$$

is the ML estimator of $\boldsymbol{\Sigma}$.

Exercise 5.4 Consider a binary classification task with the data in $\mathbb{R}$. Let $x_1 = +1$ and $x_2 = +3$ belong to the first class, and $x_3 = -2$ and $x_4 = -6$ to the second class.

(a) Determine the Gaussian ML discriminant rule by first estimating the mean values and covariances.
(b) Determine Fisher's discriminant rule for the same problem. Compare the result with part (a).

Exercise 5.5 Consider a multi-class classification task with the data in $\mathbb{R}$, given as:

- First class: $x_1 = +1$ and $x_2 = +3$
- Second class: $x_3 = -2$ and $x_4 = -4$
- Third class: $x_5 = -6$ and $x_6 = -8$

(a) Show that all classes have the same covariance. What is the Gaussian ML discriminant rule?
(b) Determine Fisher's discriminant rule and compare the result with part (a).

Exercise 5.6 Let $\boldsymbol{\Delta} = \left(\delta_{ij}\right)_{i,j=1,\ldots,n}$ be a dissimilarity matrix between n objects $v_1, \ldots, v_n$, which additionally satisfies the *ultrametric inequality*

$$\delta_{ij} \le \max\{\delta_{ik}, \delta_{kj}\} \text{ for all } i, j, k = 1 \ldots, n .$$

For any $q \ge 0$ define the sets

$$\mathcal{S}_j^{(q)} = \{i \mid \delta_{ij} \le q\}, \ \ j = 1, \ldots, n .$$

(a) Show that $\mathcal{S}_i^{(q)}$ and $\mathcal{S}_j^{(q)}$ are either disjoint or identical sets. Furthermore, show that $\bigcup_{i=1}^{n} \mathcal{S}_i^{(q)} = \{1, \ldots, n\}$. Hence, by selecting appropriate representatives a clustering is created where the dissimilarity between objects within each cluster is less than or equal to q and between objects in different clusters is greater than q.
(b) For which values of q does a single cluster or clusters of single elements occur?
(c) Represent an ultrametric dissimilarity matrix by a binary tree (a dendrogram) reflecting the cluster process at different levels of q.

Chapter 6
Support Vector Machines

In 1992, Boser, Guyon and Vapnik [7] introduced a supervised algorithm for classification that after numerous extensions is now known as *Support Vector Machines* (SVMs). Support Vector Machines denotes a class of algorithms for classification and regression, which represent the current state of the art. The algorithm determines a small subset of points—the support vectors—in a Euclidean space such that a hyperplane determined solely by these vectors separates two large classes of points at its best. The purpose of this chapter is to introduce the key methodology based on convex optimization and kernel functions.

Support vector machines are a supervised learning method for classification, regression and outlier detection. It is among the best "off-the-shelf" methods for this purpose. By the use of kernels, it is highly flexible and appropriate for many practical situations. There are efficient implementations from convex optimization and specially tailored algorithm like the *sequential minimal optimization (SMO)*. SVMs are particularly powerful in high dimensions and often yield amazingly good results even if the number of dimensions is greater than the number of samples. The main advantage is that once the algorithm is trained the decision rule depends on a small number of support vectors and can be efficiently computed for new observations.

We first start with a representation of hyperplanes and the distance between parallel hyperplanes in $\mathbb{R}^p$, the p-dimensional Euclidean space.

6.1 Hyperplanes and Margins

Given a vector $\mathbf{a} \in \mathbb{R}^p$, the p-dimensional Euclidean space with norm $\|\mathbf{x}\| = \sqrt{\sum_{i=1}^p x_i^2}$, $\mathbf{x} = (x_1, \ldots, x_p)^T$. The set of points in $\mathbb{R}^p$ which are orthogonal to $\mathbf{a}$, $\mathbf{a} \neq \mathbf{0}$, forms a $(p-1)$-dimensional linear space

R. Mathar et al., *Fundamentals of Data Analytics*,
https://doi.org/10.1007/978-3-030-56831-3_6

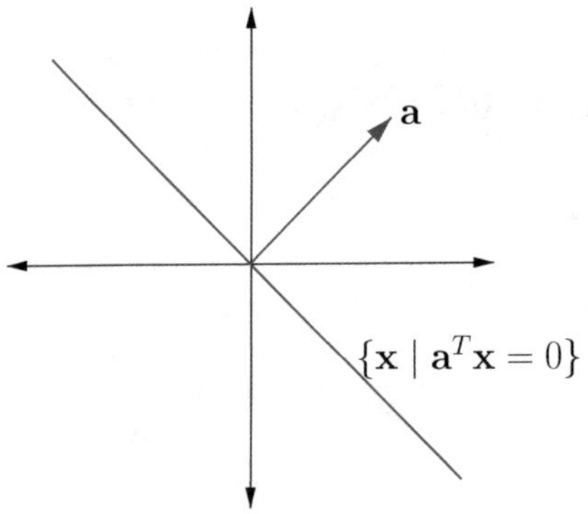

Fig. 6.1 The one-dimensional linear subspace orthogonal to **a** in two dimensions

$$\mathcal{L} = \{\mathbf{x} \in \mathbb{R}^p \mid \mathbf{a}^T\mathbf{x} = 0\}. \tag{6.1}$$

A corresponding picture in two dimensions is given in Fig. 6.1.

We now want to shift the linear space $\mathcal{L}$ by a certain amount in the direction of **a**. For this purpose let $b \in \mathbb{R}$ be a scalar and define the set

$$\mathcal{H} = \{\mathbf{x} \in \mathbb{R}^p \mid \mathbf{a}^T\mathbf{x} - b = 0\}. \tag{6.2}$$

$\mathcal{H}$ is called a $(p-1)$-dimensional *hyperplane*.

Since $\mathbf{a}^T\mathbf{a} = \|\mathbf{a}\|^2$ the following holds

$$\mathbf{a}^T\mathbf{x} - b = 0 \Leftrightarrow \mathbf{a}^T\mathbf{x} - \frac{\mathbf{a}^T\mathbf{a}}{\|\mathbf{a}\|^2}b = 0 \Leftrightarrow \mathbf{a}^T\left(\mathbf{x} - \frac{b}{\|\mathbf{a}\|^2}\mathbf{a}\right) = 0. \tag{6.3}$$

Hence, $\mathcal{H}$ is a $(p-1)$-dimensional linear space shifted by $\frac{b}{\|\mathbf{a}\|^2}\mathbf{a}$ in the direction of **a**.

Note that if **a** has length one, i.e., $\|\mathbf{a}\| = 1$ then b is the amount by which the linear space is shifted from the origin. Figure 6.2 depicts the situation in two dimensions.

A hyperplane separates the underlying space into two half-spaces, namely the set of points lying on one side of a hyperplane or the other. Both half-spaces can be taken as closed, hence containing the hyperplane, or open with the hyperplane excluded from the half-space.

$$\mathcal{G} = \{\mathbf{x} \in \mathbb{R}^p \mid \mathbf{a}^T\mathbf{x} - b \leq 0\}$$

is called a (closed) half-space. Substituting "$\leq$" by "$\geq$" yields the complementary half-space on the other side of the hyperplane.

The next question concerns the distance between two parallel hyperplanes

$$\mathcal{H}_1 = \{\mathbf{x} \in \mathbb{R}^p \mid \mathbf{a}^T\mathbf{x} - b_1 = 0\} \text{ and } \mathcal{H}_2 = \{\mathbf{x} \in \mathbb{R}^p \mid \mathbf{a}^T\mathbf{x} - b_2 = 0\}.$$

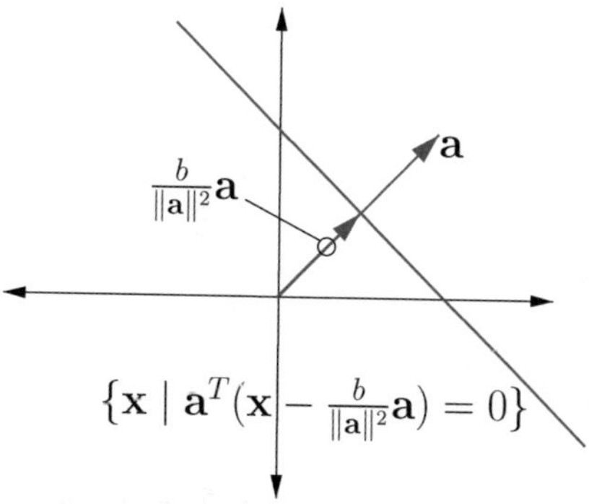

Fig. 6.2 The one-dimensional hyperplane orthogonal to **a**, shifted by $\frac{b}{\|\mathbf{a}\|^2}\mathbf{a}$ from the origin

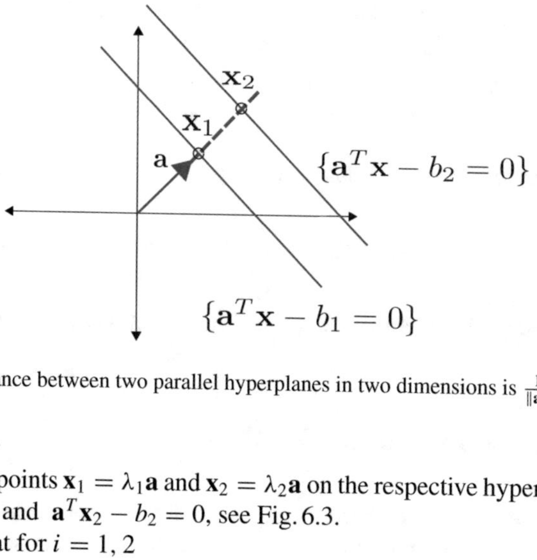

Fig. 6.3 The distance between two parallel hyperplanes in two dimensions is $\frac{1}{\|\mathbf{a}\|}|b_2 - b_1|$

First select two points $\mathbf{x}_1 = \lambda_1\mathbf{a}$ and $\mathbf{x}_2 = \lambda_2\mathbf{a}$ on the respective hyperplane such that $\mathbf{a}^T\mathbf{x}_1 - b_1 = 0$ and $\mathbf{a}^T\mathbf{x}_2 - b_2 = 0$, see Fig. 6.3.

It follows that for $i = 1, 2$

$$\lambda_i\mathbf{a}^T\mathbf{a} - b_i = 0 \Leftrightarrow \lambda_i\|\mathbf{a}\|^2 - b_i = 0 \Leftrightarrow \lambda_i = b_i/\|\mathbf{a}\|^2,$$

and

$$\begin{aligned}\|\mathbf{x}_2 - \mathbf{x}_1\| &= \|\lambda_2\mathbf{a} - \lambda_1\mathbf{a}\| = |\lambda_2 - \lambda_1|\|\mathbf{a}\| \\ &= \left|\frac{b_2}{\|\mathbf{a}\|^2} - \frac{b_1}{\|\mathbf{a}\|^2}\right|\|\mathbf{a}\| = \frac{1}{\|\mathbf{a}\|}|b_2 - b_1|.\end{aligned}$$

Hence, the distance between the hyperplanes $\mathcal{H}_1$ and $\mathcal{H}_2$ is

$$d(\mathcal{H}_1, \mathcal{H}_2) = \frac{1}{\|\mathbf{a}\|}|b_2 - b_1|. \tag{6.4}$$

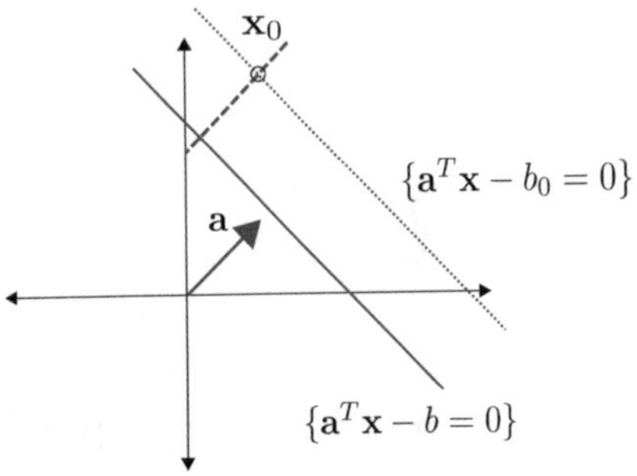

Fig. 6.4 The minimum distance between a point $\mathbf{x}_0$ and a hyperplane is $\frac{1}{\|\mathbf{a}\|}|b - \mathbf{a}^T\mathbf{x}_0|$

Finally, we determine the minimum distance between a given point $\mathbf{x}_0 \in \mathbb{R}^p$ and a hyperplane $\mathcal{H} = \{\mathbf{x} \mid \mathbf{a}^T\mathbf{x} - b = 0\}$. Consider the auxiliary hyperplane parallel to $\mathcal{H}$ and containing $\mathbf{x}_0$.

$$\mathcal{H}_0 = \{\mathbf{x} \mid \mathbf{a}^T\mathbf{x} - \mathbf{a}^T\mathbf{x}_0 = 0\}.$$

By (6.4) the distance between $\mathcal{H}$ and $\mathcal{H}_0$, hence, the minimum distance between $\mathcal{H}$ and $\mathbf{x}_0$ is

$$d(\mathcal{H}, \mathbf{x}_0) = \frac{1}{\|\mathbf{a}\|}|b - \mathbf{a}^T\mathbf{x}_0|.$$

This distance is called the *margin* of $\mathbf{x}_0$, see Fig. 6.4 for a two-dimensional picture.

6.2 The Optimum Margin Classifier

The above derivations will be used to introduce the concept of optimal separating hyperplanes. Given a training set of n observations from two classes

$$(\mathbf{x}_1, y_1), \ldots, (\mathbf{x}_n, y_n), \ \mathbf{x}_i \in \mathbb{R}^p, \ y_i \in \{-1, 1\}.$$

$\mathbf{x}_i$ denotes the observed value and y_i the class label denoted by -1 or $+1$. We first assume that there exists a separating hyperplane between the classes

$$\mathcal{H} = \{\mathbf{x} \in \mathbb{R}^p \mid \mathbf{a}^T\mathbf{x} + b = 0\}$$

for some $\mathbf{a} \in \mathbb{R}^p$ and some scalar $b \in \mathbb{R}$. Of course the separating hyperplane may not be unique so that we have different choices for $\mathbf{a}$ and b, see Fig. 6.5 for an example in $\mathbb{R}^2$. Intuitively it is clear that the black line is the optimum choice for a separating hyperplan. It has the property that the minimum distance to any of the points in both classes is maximized. To formalize this objective in a proper way we observe that

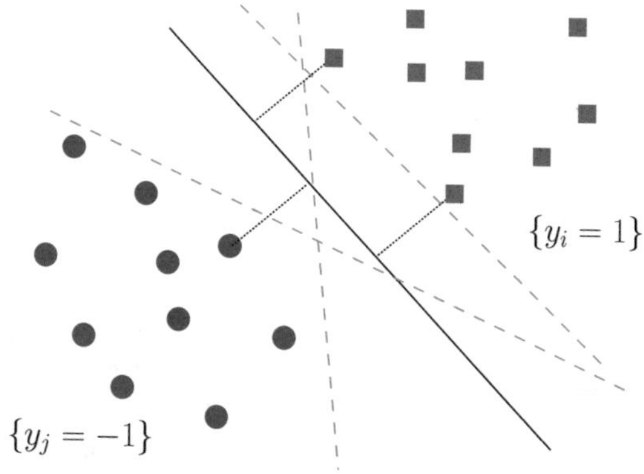

Fig. 6.5 Different separating hyperplanes for a two-class problem in $\mathbb{R}^2$. Which one is optimal?

for any separating hyperplane there exists some $\gamma \geq 0$ such that for all $i = 1, \ldots, n$

$$y_i = +1 \Rightarrow \mathbf{a}^T \mathbf{x}_i + b \geq \gamma$$
$$y_i = -1 \Rightarrow \mathbf{a}^T \mathbf{x}_i + b \leq -\gamma$$

Hence, for some $\gamma \geq 0$ it holds that

$$y_i(\mathbf{a}^T \mathbf{x}_i + b) \geq \gamma \quad \text{for all} \quad i = 1, \ldots, n.$$

The objective of finding a hyperplane which maximizes the minimum distance to any of the points can now be formalized as

$$\begin{aligned} &\max_{\gamma \geq 0, \mathbf{a} \in \mathbb{R}^p, b \in \mathbb{R}} \frac{\gamma}{\|\mathbf{a}\|} \\ &\text{such that } y_i(\mathbf{a}^T \mathbf{x}_i + b) \geq \gamma, \quad i = 1, \ldots, n. \end{aligned} \tag{6.5}$$

This problem is not scale invariant. With γ, $\mathbf{A}$, b any multiple $u\gamma$, $u\mathbf{A}$, ub, $u > 0$, is also a solution. The problem is rendered scale invariant by dividing both sides of (6.5) by $\gamma > 0$. Instead of maximizing the quotient we can also minimize its reciprocal to obtain

$$\begin{aligned} &\min_{\gamma \geq 0, \mathbf{a} \in \mathbb{R}^p, b \in \mathbb{R}} \left\| \frac{\mathbf{a}}{\gamma} \right\| \\ &\text{such that } y_i\Big(\frac{\mathbf{a}^T}{\gamma}\mathbf{x}_i + \frac{b}{\gamma}\Big) \geq 1 \quad i = 1, \ldots, n. \end{aligned}$$

Obviously the solution space does not change by replacing $\mathbf{a}/\gamma$ and b/γ by $\mathbf{a}$ and b, respectively, which yields

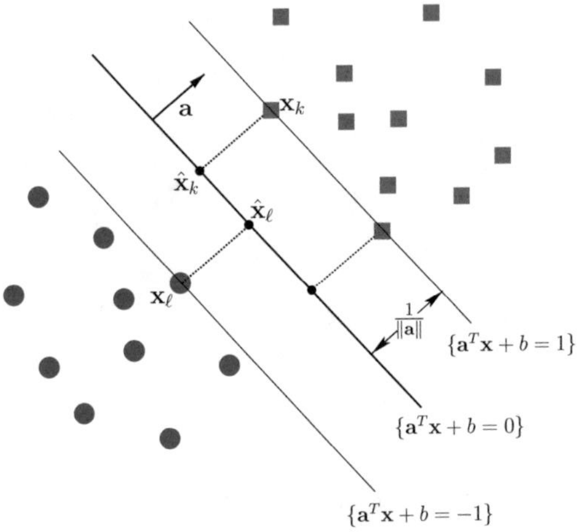

Fig. 6.6 The solution to the optimum margin classifier problem

$$\min_{\mathbf{a}\in\mathbb{R}^p, b\in\mathbb{R}} \quad \|\mathbf{a}\|$$
$$\text{such that } y_i\big(\mathbf{a}^T\mathbf{x}_i + b\big) \geq 1 \quad i = 1, \ldots, n.$$

The objective function can be modified by squaring the norm and multiplying by the factor $\frac{1}{2}$, both monotonic transformations which do not change the solution. Finally, the *optimum margin classifier* problem (OMC) reads as

$$\max_{\mathbf{a}\in\mathbb{R}^p, b\in\mathbb{R}} \quad \frac{1}{2}\|\mathbf{a}\|^2 \tag{6.6}$$
$$\text{such that } y_i(\mathbf{a}^T\mathbf{x}_i + b) \geq 1, \quad i = 1, \ldots, n.$$

The above approach is graphically depicted in Fig. 6.6. The distance of the hyperplane from the nearest points to each side is $\frac{1}{\|\mathbf{a}\|}$ as is known from (6.4) in Sect. 6.1. From the picture we can also see that only a few points determine the optimum margin hyperplane, two in the red class and a single one in the blue class. These points are called *support vectors*. They satisfy the constraints in (6.6) with equality.

Note that $\mathbf{x}_i \in \mathbb{R}^p$ and $y_i \in \{-1, 1\}$ are given by the training set and the only variables are $\mathbf{a} \in \mathbb{R}^p$ and $b \in \mathbb{R}$. The factor $1/2$ is used to obtain the problem in standard form of quadratic optimization.

The OMC problem (6.6) is a quadratic optimization problem with linear constraints. It is a special case of convex optimization. There exist efficient algorithms for finding a solution in very short time even for huge problem sizes. Commercial or public domain generic software can be used to solve the problem numerically.

Assume that $\mathbf{a}^*$ is an optimum solution of (6.6) and $\mathbf{x}_k$ is a support vector with maximum margin fulfilling the corresponding constraint in (6.6) with equality. Hence,

$$y_k\left(\mathbf{a}^{*T}\mathbf{x}_k + b\right) = 1$$

so that

$$b^* = y_k - \mathbf{a}^{*T}\mathbf{x}_k \tag{6.7}$$

holds for the optimal b. The solution $(\mathbf{a}^*, b^*)$ is called the *optimal margin classifier*.

By the above, the OMC problem is solved. However, in certain cases we can do better by considering the dual problem. This will also lead us to *kernels*, thus making the approach extremely flexible and also appropriate for cases when a separating hyperplane does not exist.

6.3 Convex Optimization and Duality

Quadratic optimization problems with linear constraints can be easily transformed into a so called dual problem, as will be done for support vector machines in the following. The following is a short introduction to Lagrange duality theory. A good reference to an in-depth treatment is the book of Boyd and Vandenberghe [8].

A convex optimization problem is described by a convex objective function and convex, partially even linear constraints.

$$\begin{aligned} &\text{minimize } f_0(\mathbf{x}) \\ &\text{such that } f_i(\mathbf{x}) \le 0,\ i = 1, \dots, m \\ &\qquad\qquad h_i(\mathbf{x}) = 0,\ i = 1, \dots, r \end{aligned} \tag{6.8}$$

where $f_0, f_i : \mathbb{R}^p \to \mathbb{R}$ are convex functions and $h_i : \mathbb{R}^p \to \mathbb{R}$ are linear functions.

The *Lagrangian prime function* includes dual parameters $\boldsymbol{\lambda} = (\lambda_1, \dots, \lambda_m)$ and $\boldsymbol{\nu} = (\nu_1, \dots, \nu_r)$ as follows

$$L(\mathbf{x}, \boldsymbol{\lambda}, \boldsymbol{\nu}) = f_0(\mathbf{x}) + \sum_{i=1}^{m} \lambda_i f_i(\mathbf{x}) + \sum_{i=1}^{r} \nu_i h_i(\mathbf{x}).$$

From this the *Lagrangian dual function* is derived as

$$g(\boldsymbol{\lambda}, \boldsymbol{\nu}) = \inf_{\mathbf{x} \in \mathcal{D}} L(\mathbf{x}, \boldsymbol{\lambda}, \boldsymbol{\nu}).$$

The set $\mathcal{D} = \bigcap_{i=0}^{m} \operatorname{dom}(f_i) \cap \bigcap_{i=1}^{r} \operatorname{dom}(h_i)$ is the domain where all functions are well defined.

Since the dual function is linear in $\boldsymbol{\lambda}$ and $\boldsymbol{\nu}$ and since the infimum over linear functions is concave, the Lagrangian dual function (6.3) is always a concave function, even if f_i and h_i are not convex and linear, respectively. Having the dual function at hand the *Lagrange dual problem* is defined by

$$\begin{aligned} &\text{maximize } g(\boldsymbol{\lambda}, \boldsymbol{\nu}) \\ &\text{such that } \boldsymbol{\lambda} \geq \mathbf{0} \end{aligned} \tag{6.9}$$

where "$\geq$" is to be understood component-wise.

A central theorem is the so called *weak duality theorem* which states that

$$g(\boldsymbol{\lambda}^*, \boldsymbol{\nu}^*) \leq f_0(\mathbf{x}^*)$$

$\boldsymbol{\lambda}^*, \boldsymbol{\nu}^*$ denoting the optimal solution of the dual problem (6.8) and $\mathbf{x}^*$ the optimal solution of the primal problem (6.8).

Strong duality states that both solutions are equal

$$g(\boldsymbol{\lambda}^*, \boldsymbol{\nu}^*) = f_0(\mathbf{x}^*). \tag{6.10}$$

Strong duality entails that solving the dual problem leads to the same solution as the primal one. Hence the primal can be substituted by the dual problem. This is especially valuable if the primal problem is complicated and the dual allows for a fast and easy solution.

The difference $f_0(\mathbf{x}^*) - g(\boldsymbol{\lambda}^*, \boldsymbol{\nu}^*)$ is called *duality gap*. Hence, if strong duality holds then the duality gap is zero.

Strong duality implies that

$$\begin{aligned} f_0(\mathbf{x}^*) &= g(\boldsymbol{\lambda}^*, \boldsymbol{\nu}^*) \\ &= \inf_{\mathbf{x}\in\mathcal{D}} \left(f_0(\mathbf{x}) + \sum_{i=1}^{m} \lambda_i^* f_i(\mathbf{x}) + \sum_{i=1}^{r} \nu_i^* h_i(\mathbf{x}) \right) \\ &\leq f_0(\mathbf{x}^*) + \sum_{i=1}^{m} \lambda_i^* f_i(\mathbf{x}^*) + \sum_{i=1}^{r} \nu_i^* h_i(\mathbf{x}^*) \\ &\leq f_0(\mathbf{x}^*). \end{aligned}$$

Since for the optimal solutions $\lambda_i^* \geq 0$, $f_i(\mathbf{x}^*) \leq 0$ and $h_i(\mathbf{x}^*) = 0$ holds, the so called *complementary slackness condition*

$$\sum_{i=1}^{m} \lambda_i^* f_i(\mathbf{x}^*) = 0 \tag{6.11}$$

follows.

The so called *Slater's conditions* ensure that strong duality (6.10) holds, cf. [8], Sect. 5.2.3. Furthermore, if the constraints are purely linear, then Slater's conditions are satisfied, which will be the case for the SVM problem (6.6).

If additionally $f_i, i = 0, \ldots, m$, are differentiable and $\mathbf{x}^*$ is an inner point, then

$$\nabla_{\mathbf{x}} L(\mathbf{x}^*, \boldsymbol{\lambda}^*, \boldsymbol{\nu}^*) = 0.$$

Putting things together we end up with the following theorem.

Theorem 6.1 *If for a convex optimization problem (6.8) Slater's conditions hold and f_i, $i = 0, \ldots, m$, are differentiable, then $\mathbf{x}^*$ and $(\boldsymbol{\lambda}^*, \boldsymbol{\nu}^*)$ are primal and dual optimal if and only if*

$$f_i(\mathbf{x}^*) \le 0, \; i = 1, \ldots, m, \; h_i(\mathbf{x}^*) = 0, \; i = 1, \ldots, r \tag{6.12}$$
$$\boldsymbol{\lambda}^* \ge 0 \tag{6.13}$$
$$\lambda_i^* f_i(\mathbf{x}^*) = 0, \; i = 1, \ldots, m \tag{6.14}$$
$$\nabla_{\mathbf{x}} L(\mathbf{x}^*, \boldsymbol{\lambda}^*, \boldsymbol{\nu}^*) = 0 \tag{6.15}$$

Equations (6.12)–(6.15) are called the *Karush-Kuhn-Tucker conditions*. The first equation (6.12) ensures that the primal constraints are satisfied, (6.13) does the same for the dual constraints. Complementary slackness is required by (6.14) and primal optimality follows from (6.15).

The value of the KKT conditions lies in guessing a solution $\mathbf{x}^*$ and then verifying that it is actually optimal. Normally, computing the partial derivatives (6.15) gives a good hint to what the optimal solution would look like.

6.4 Support Vector Machines and Duality

We now apply the previously developed theory to support vector machines. Recall that a data set $\{(\mathbf{x}_1, y_1), \ldots, (\mathbf{x}_n, y_n)\}$, $\mathbf{x}_i \in \mathbb{R}^p$, $y_i \in \{-1, 1\}$ is given. The OMC problem (6.6) is described by

$$\max_{\mathbf{a}\in\mathbb{R}^p, b\in\mathbb{R}} \frac{1}{2}\|\mathbf{a}\|^2$$
$$\text{such that } y_i(\mathbf{a}^T\mathbf{x}_i + b) \ge 1, \quad i = 1, \ldots, n.$$

It consists of a convex objective function and linear constraints only. The Lagrangian reads as

$$L(\mathbf{a}, b, \boldsymbol{\lambda}) = \frac{1}{2}\|\mathbf{a}\|^2 - \sum_{i=1}^{n} \lambda_i \big(y_i(\mathbf{a}^T\mathbf{x}_i + b) - 1\big)$$

Setting the partial derivatives w.r.t. **a** equal to 0

$$\nabla_{\mathbf{a}} L(\mathbf{a}, b, \boldsymbol{\lambda}) = \mathbf{a} - \sum_{i=1}^{n} \lambda_i y_i \mathbf{x}_i = 0$$

yields the optimum solution

$$\mathbf{a}^* = \sum_{i=1}^{n} \lambda_i y_i \mathbf{x}_i. \tag{6.16}$$

The derivative w.r.t. b is

$$\frac{d}{db} L(\mathbf{a}, b, \boldsymbol{\lambda}) = \sum_{i=1}^{n} \lambda_i y_i$$

yielding for the optimal dual solution

$$\sum_{i=1}^{n} \lambda_i^* y_i = 0. \tag{6.17}$$

The Lagrangian dual function has the form

$$\begin{aligned}
g(\boldsymbol{\lambda}) &= L(\mathbf{a}^*, b^*, \boldsymbol{\lambda}) \\
&= \frac{1}{2}\|\mathbf{a}^*\|^2 - \sum_{i=1}^{n} \lambda_i \big(y_i(\mathbf{a}^{*T}\mathbf{x}_i + b^*) - 1\big) \\
&= \sum_{i=1}^{n} \lambda_i + \frac{1}{2}\Big(\sum_{i=1}^{n} \lambda_i y_i\Big)^T \Big(\sum_{i=1}^{n} \lambda_i y_i\Big) \\
&\quad - \Big(\sum_{i=1}^{n} \lambda_i y_i\Big)\Big(\sum_{j=1}^{n} \lambda_j y_j \mathbf{x}_j\Big)^T \mathbf{x}_i - \sum_{i=1}^{n} \lambda_i y_i b^* \\
&= \sum_{i=1}^{n} \lambda_i - \frac{1}{2} \sum_{i,j=1}^{n} y_i y_j \lambda_i \lambda_j \mathbf{x}_i^T \mathbf{x}_j - \sum_{i=1}^{n} \lambda_i y_i b^*.
\end{aligned}$$

Since by (6.17) the optimum $\boldsymbol{\lambda}^*$ satisfies $\sum_{i=1}^{n} \lambda_i^* y_i = 0$, the Lagrangian dual problem for SVMs reduces to

$$\begin{aligned}
&\max_{\lambda} \Big\{ \sum_{i=1}^{n} \lambda_i - \frac{1}{2} \sum_{i,j=1}^{n} y_i y_j \lambda_i \lambda_j \mathbf{x}_i^T \mathbf{x}_j \Big\} \\
&\text{such that } \lambda_i \geq 0,\ i = 1, \ldots, n, \\
&\sum_{i=1}^{n} \lambda_i y_i = 0.
\end{aligned} \tag{6.18}$$

Note that the variables in (6.18) are $\lambda_1, \ldots, \lambda_n$ and that there are $n + 1$ constraints, n denoting the number of data points. The objective function is quadratic which allows the problem to be solved efficiently by appropriate convex programming software.

The primal problem (6.6) has $p + 1$ variables and n constraints. It depends on the size of n and p which problem can be solved more efficiently.

Let λ_i^* denote the optimum solution of (6.18). From (6.16) we conclude that

$$\mathbf{a}^* = \sum_{i=1}^{n} \lambda_i^* y_i \mathbf{x}_i \tag{6.19}$$

and from (6.7) that

$$b^* = y_k - \mathbf{a}^{*T} \mathbf{x}_k$$

for some support vector $\mathbf{x}_k$, which leaves us with the problem of determining the support vectors.

Slater's conditions are satisfied, as pointed out earlier. Strong duality holds and complementary slackness follows from the KKT conditions (6.14), i.e.,

$$\lambda_i^* \big(y_i (\mathbf{a}^{*T} \mathbf{x}_i + b^*) - 1 \big) = 0, \quad i = 1, \ldots, n.$$

Hence,

$$\begin{aligned} \lambda_i^* > 0 &\Rightarrow y_i (\mathbf{a}^{*T} \mathbf{x}_i + b^*) = 1 \\ y_i (\mathbf{a}^{*T} \mathbf{x}_i + b^*) > 1 &\Rightarrow \lambda_i^* = 0 \end{aligned}$$

This shows that support vectors $\mathbf{x}_k$ are characterized by a corresponding $\lambda_k^* = 0$. They are the ones which fulfill the primal constraints with equality and hence yield the maximum margin. This also shows that in (6.19) only addends with nonzero λ_i^* need to be included, which are normally much fewer than n. To formalize this idea let

$$\mathcal{S} = \{ i \mid 1 \le i \le n, \lambda_i^* > 0 \}.$$

The optimum margin hyperplane is then determined by

$$\mathbf{a}^* = \sum_{i \in \mathcal{S}} \lambda_i^* y_i \mathbf{x}_i \text{ and } b^* = y_k - \mathbf{a}^{*T} \mathbf{x}_k \text{ for some } k \in \mathcal{S}. \tag{6.20}$$

Once a support vector machine is trained, i.e., $\boldsymbol{\lambda}^*, \mathbf{a}^*, b^*$ are computed by the training set $\{(\mathbf{x}_1, y_1), \ldots, (\mathbf{x}_n, y_n)\}$, a newly observed data $\mathbf{x}$ is classified by the decision rule

$$d(\mathbf{x}) = \mathbf{a}^{*T} \mathbf{x} + b^* = \Big(\sum_{i \in \mathcal{S}} \lambda_i^* y_i \mathbf{x}_i \Big)^T \mathbf{x} + b^* = \sum_{i \in \mathcal{S}} \lambda_i^* y_i \mathbf{x}_i^T \mathbf{x} + b^* \lesseqgtr 0 \, .$$

If $d(\mathbf{x}) \ge 0$ decide that $\mathbf{x}$ belongs to class labeled $y = 1$, otherwise decide that the class is $y = -1$.

Note that the cardinality of $\mathcal{S}$ is normally much less than n, which makes the decision rule very efficient and only dependent on the support vectors.

Moreover, the decision rule depends merely on the inner products $\mathbf{x}_i^T\mathbf{x}$, $i \in \mathcal{S}$. This allows for generalizing support vector machines to nonlinear situations where no separating hyperplane exists.

6.5 Non-Separability and Robustness

We next deal with the case that no separating hyperplane exists. Applying the approach of the previous section will not work, since the set of feasible solution is empty. Running convex optimization software will end up with exactly this reply. Nevertheless there may be situations that only a few data is on the "wrong side" of a certain hyperplane like shown in Fig. 6.7. The problem must be formulated as an optimization problem that allows for misclassified data or data lying in the margin area.

Another reason for modifying the original problem is to make the solution robust against outliers. Figure 6.8 demonstrates a situation where one additional data causes a drastic swing of the separating hyperplane yielding a solution not adequate for the main mass of data.

Both problems are addressed by using ℓ_1-regularization. Recall that the training set is $(\mathbf{x}_1, y_1), \ldots, (\mathbf{x}_n, y_n)$, $\mathbf{x}_i \in \mathbb{R}^p$, with class labels $y_i \in \{-1, 1\}$.We reformulate the original optimization (6.6) as follows:

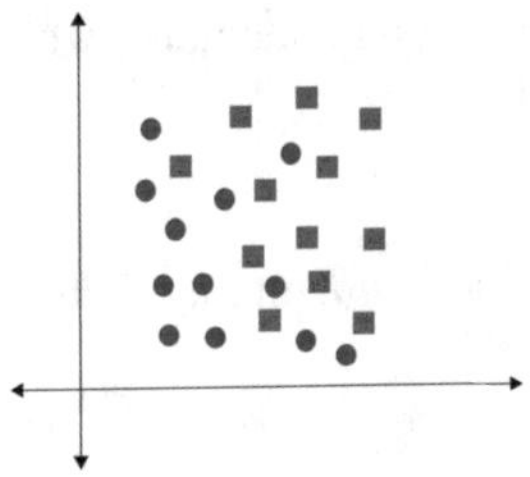

Fig. 6.7 There is no separating hyperplane although essentially two classes are recognizable

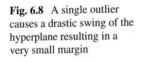

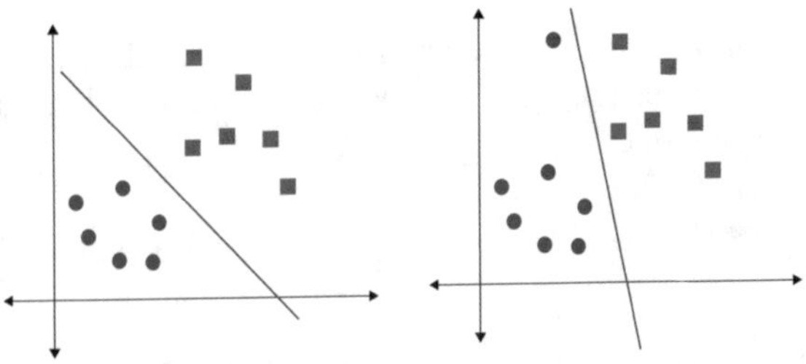

Fig. 6.8 A single outlier causes a drastic swing of the hyperplane resulting in a very small margin

$$\begin{aligned} &\min_{\mathbf{a},b} \frac{1}{2}\|\mathbf{a}\|^2 + c\sum_{i=1}^{n} \xi_i \\ &\text{such that } y_i(\mathbf{a}_i^T\mathbf{x}_i + b) \geq 1 - \xi_i \\ &\xi_i \geq 0, \quad i = 1, \ldots, n \end{aligned} \tag{6.21}$$

Assume that $\mathbf{a}, b$ and ξ_i is a feasible solution of (6.21). If $y_i(\mathbf{a}^T\mathbf{x}_i + b) = 1 - \xi_i$, $\xi_i > 0$, then a cost of $c\xi_i$ is incurred. Parameter c controls the weight of the additional costs in the objective function. If c is large, the influence of $\sum_i \xi_i$ becomes prevailing so that most data will have margin of atleast one.

Furthermore, if $\xi_i > 1$ then the corresponding data $\mathbf{x}_i$ is misclassified. On the other hand, if $0 < \xi_i \leq 1$ then the data is correctly classified but may lie in the margin.

Let $x^+ = \max\{x, 0\}$ denote the *positive part* of real number $x \in \mathbb{R}$. The constraints $y_i(\mathbf{a}_i^T\mathbf{x}_i + b) \geq 1 - \xi_i$ may be rewritten as

$$\xi_i \geq \left[1 - y_i(\mathbf{a}_i^T\mathbf{x}_i + b)\right]^+$$

so that jointly with the constraints $\xi_i \geq 0$ optimization (6.21) has an equivalent unconstrained formulation, namely

$$\min_{\mathbf{a},b} \frac{1}{2}\|\mathbf{a}\|^2 + c\sum_{i=1}^{n} \left[1 - y_i(\mathbf{a}_i^T\mathbf{x}_i + b)\right]^+.$$

This representation lets us conclude as follows.

$$\begin{aligned} y_i(\mathbf{a}_i^T\mathbf{x}_i + b) > 1 &\Rightarrow \text{ there is no loss } \mathbf{x}_i \text{ lies outside margin} \\ y_i(\mathbf{a}_i^T\mathbf{x}_i + b) = 1 &\Rightarrow \text{ there is no loss, } \mathbf{x}_i \text{ lies on the margin} \\ y_i(\mathbf{a}_i^T\mathbf{x}_i + b) < 1 &\Rightarrow \text{ contribution to loss, } \mathbf{x}_i \text{ violates the margin constraint} \end{aligned}$$

We will now consider the dual problem of (6.21). The first step is to determine the Lagrangian prime function which reads as

$$L(\mathbf{a}, b, \boldsymbol{\xi}, \boldsymbol{\lambda}, \boldsymbol{\gamma}) = \frac{1}{2}\|\mathbf{a}\|^2 + c\sum_{i=1}^{n} \xi_i - \sum_{i=1}^{n} \lambda_i\big(y_i(\mathbf{a}^T\mathbf{x}_i + b) - 1 + \xi_i\big) - \sum_{i=1}^{n} \gamma_i\xi_i$$

with the obvious notation $\boldsymbol{\xi} = (\xi_1, \ldots, \xi_n)$ and $\boldsymbol{\gamma} = (\gamma_1, \ldots, \gamma_n)$. $\boldsymbol{\lambda}$ and $\boldsymbol{\gamma}$ are the Lagrangian multipliers.

Following the same steps as above the dual problem is obtained as

$$\max_{\boldsymbol{\lambda}} \Big\{ \sum_{i=1}^{n} \lambda_i - \frac{1}{2} \sum_{i,j=1}^{n} y_i y_j \lambda_i \lambda_j \mathbf{x}_i^T \mathbf{x}_j \Big\}$$

$$\text{such that } 0 \le \lambda_i \le c, \quad i = 1, \ldots, n \tag{6.22}$$

$$\sum_{i=1}^{n} \lambda_i y_i = 0$$

The only difference to (6.18) are the additional constraints $\lambda_i \le c, i = 1, \ldots, n$.

Let $\boldsymbol{\lambda}^*$ be an optimal solution of (6.22) and let

$$\mathcal{S} = \{1 \le i \le n, \mid \lambda_i^* > 0\}.$$

Then

$$\mathbf{a}^* = \sum_{i \in \mathcal{S}} \lambda_i^* y_i \mathbf{x}_i \tag{6.23}$$

is the optimum **a**. Complementary slackness conditions yield

$$0 < \lambda_i^* < c \Rightarrow y_i(\mathbf{a}^{*T} \mathbf{x}_i + b^*) = 1 \quad \text{(margin SVs)} \tag{6.24}$$

$$\lambda_i^* = c \Rightarrow y_i(\mathbf{a}^{*T} \mathbf{x}_i + b^*) \le 1 \quad \text{(SVs with margin errors)} \tag{6.25}$$

$$\lambda_i^* = 0 \Rightarrow y_i(\mathbf{a}^{*T} \mathbf{x}_i + b^*) \ge 1 \quad \text{(non SVs)} \tag{6.26}$$

The points with property (6.24) are support vectors determining the hyperplane. Points which satisfy (6.25) are support vectors with margin errors. Condition (6.26) identifies points which are no support vectors but are correctly classified. Figure 6.9 visualizes the possible location and corresponding λ-values of data.

For any support vectors, $\mathbf{x}_k$ say, satisfying (6.24)

$$b^* = y_k - \mathbf{a}^{*T} \mathbf{x}_k \tag{6.27}$$

is the optimal b^*.

With the optimal $\mathbf{a}^*$ from (6.23) and b^* from (6.27) a newly observed point $\mathbf{x} \in \mathbb{R}^p$ is classified by first computing

$$d(\mathbf{x}) = \mathbf{a}^{*T} \mathbf{x} + b^* = \Big(\sum_{i \in \mathcal{S}} \lambda_i^* y_i \mathbf{x}_i \Big)^T \mathbf{x} + b^* = \sum_{i \in \mathcal{S}} \lambda_i^* y_i \mathbf{x}_i^T \mathbf{x} + b^*. \tag{6.28}$$

For a *hard classifier* the final decision is

$$\text{Decide } y = 1 \text{ if } d(\mathbf{x}) \ge 0, \text{ otherwise decide } y = -1$$

where y denotes the class corresponding to **x**.

The *soft classifier* is

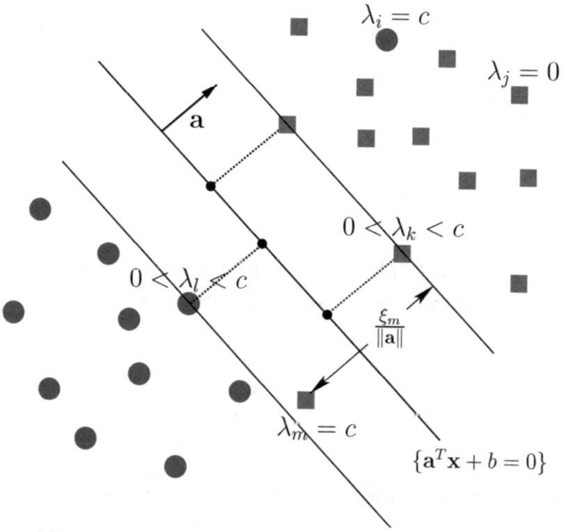

Fig. 6.9 Support vectors, correctly classified data, and margin errors with corresponding values of the dual variables λ_i

$$d_s(\mathbf{x}) = h\big(\mathbf{a}^{*T}\mathbf{x} + b^*\big)$$

with

$$h(t) = \begin{cases} -1, & \text{if } t < -1 \\ t, & \text{if } -1 \le t \le 1 \\ 1, & \text{if } t > 1 \end{cases}.$$

$d_s(\mathbf{x})$ is a real number in the interval $[-1, 1]$ representing the certainty about class correspondence.

One should note that both classifiers only depend on the inner products $\mathbf{x}_i^T\mathbf{x} = \langle \mathbf{x}_i, \mathbf{x} \rangle$ for support vectors $\mathbf{x}_i$, $i \in \mathcal{S}$.

6.6 The SMO Algorithm

An efficient heuristic algorithm for solving the dual problem (6.22) is introduced, named *sequential minimal optimization (SMO)*. We start by recalling the dual problem. Given $\mathbf{x}_1, \ldots, \mathbf{x}_n \in \mathbb{R}^p$ and class labels $y_1, \ldots, y_n \in \{-1, 1\}$,

$$\max_{\boldsymbol{\lambda}} \left\{ g(\boldsymbol{\lambda}) = \sum_{i=1}^{n} \lambda_i - \frac{1}{2} y_i y_j \lambda_i \lambda_j \mathbf{x}_i^T \mathbf{x}_j \right\}$$

$$\text{such that } 0 \le \lambda_i \le c, \quad i = 1, \ldots, n$$

$$\sum_{i=1}^{n} \lambda_i y_i = 0$$

Assume that $\boldsymbol{\lambda}$ is a feasible point, i.e., $\boldsymbol{\lambda}$ satisfied the constraints in (6.22). Cyclic coordinatewise optimization of $g(\lambda)$ does not work since

$$\lambda_j = -y_j \sum_{i \neq j}^{n} \lambda_i y_i.$$

Hence, each λ_j is determined when keeping $\lambda_i, i \neq j$, fixed.

The key idea of the SMO algorithm is to update and locally optimize two components of $\boldsymbol{\lambda}$ at the same time holding the others fixed, and cycling through this procedure.

Algorithm 1 The SMO algorithm

1: **procedure** SMO
2: **repeat**
3: 1. Select a pair (i, j) to be updated next, the one which promises the most progress
4: 2. Optimize $g(\boldsymbol{\lambda})$ w.r.t. λ_i and λ_j while keeping $\lambda_k, k \neq i, j$, fixed.
5: **until** Convergence

In step 2 $g(\boldsymbol{\lambda})$ is optimized w.r.t. to the pair λ_i and $\lambda_j, i < j$, say, with components different from i, j held fixed.

It holds that

$$\lambda_i y_i + \lambda_j y_j = - \sum_{k \neq i,j}^{n} \lambda_k y_k = r$$

with r being fixed. Since $(y_i)^2 = 1$ it follows that

$$\lambda_i = (r - \lambda_j y_j) y_i$$

and the objective function can be written

$$g(\lambda_1, \ldots, \lambda_n) = g(\lambda_1, \ldots, \lambda_{i-1}, (r - \lambda_j y_j) y_i, \lambda_{i+1}, \ldots, \lambda_j, \ldots, \lambda_n).$$

This is a quadratic function of variable λ_j of the form

$$\alpha_2 \lambda_j^2 + \alpha_1 \lambda_j + \alpha_0. \tag{6.29}$$

The maximum can be computed by differentiation. It is attained at the argument $\lambda_j = -\frac{\alpha_1}{2\alpha_2}$.

However, constraints must be obeyed to achieve a feasible solution, namely

$$0 \leq \lambda_i, \lambda_j \leq c$$
$$\lambda_j = r y_j - \lambda_i y_i y_j \tag{6.30}$$

If $y_i = y_j$, then (6.30) is simplified to

$$\lambda_j = -\lambda_i + r y_j.$$

This is a linear function which cuts the λ_j-axis at $r y_j$. It meets the box constraints $0 \leq \lambda_i, \lambda_j \leq c$ in the following cases

$$\begin{aligned} r y_j < 0 &\Rightarrow \text{ no solution} \\ 0 \leq r y_j \leq c &\Rightarrow 0 \leq \lambda_j \leq r y_j \\ c \leq r y_j \leq 2c &\Rightarrow r y_j \leq \lambda_j \leq c \end{aligned} \tag{6.31}$$

The case $r y_j > 2c$ cannot happen. See Fig. 6.10 for a graphical illustration of the case $c \leq r y_j \leq 2c$.

If $y_i \neq y_j$ then (6.30) reads as

$$\lambda_j = \lambda_i + r y_j.$$

Again we have a linear function with slope equal to 1. The box constraints are fulfilled as described by

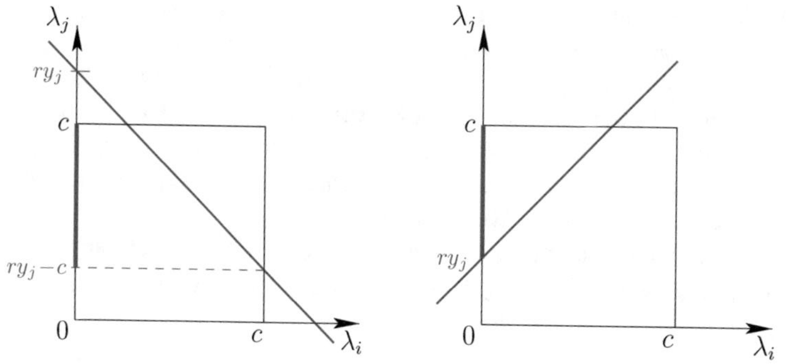

Fig. 6.10 Left: Graphical presentation of the case $y_i = y_j$ and $c \leq r y_j \leq 2c$. The linear function (blue line) has slope -1. A value for the optimum λ_j^* outside the blue bar is clipped at $r y_j - c$ or c, depending on the direction it sticks out. Right: The case $y_i \neq y_j$ and $0 \leq r y_j \leq c$. The linear function has slope $+1$. A value for the optimum λ_j^* outside the blue bar is clipped at $r y_j$ or c, depending on the direction it sticks out

$$\begin{aligned} ry_j < -c &\Rightarrow \text{no solution} \\ -c \leq ry_j \leq 0 &\Rightarrow 0 \leq \lambda_j \leq c + ry_j \\ 0 \leq ry_j \leq c &\Rightarrow ry_2 \leq \lambda_j \leq c \\ c < ry_j &\Rightarrow \text{no solution} \end{aligned} \tag{6.32}$$

If the solution λ_j^* of (6.29) exceeds the bounds, then the final solution is clipped at the corresponding bound, denoted by λ_j^{*c}.

The solution for the pairing λ_i is computed as

$$\lambda_i^{*c} = (r - \lambda_j^{*c} y_j) y_i.$$

In the case that there is no admissible solution chose another pair of indices and iterate the steps of the algorithm.

There is still potential for improvement of the algorithm by clarifying the following questions.

- What is the best choice of the next pair (i, j) to update?
- Is there an easy way to update the parameters α_k, $k = 0, 1, 2$ during execution of the algorithm.

The algorithm converges, however, an optimized choice of the follow-up pairs (i, j) may drastically speed up convergence. The algorithm was invented by John Platt in 1998, see [34]. Much more information and further improvements can be found in this paper.

6.7 Kernels for SVM

Support vector machines may also be used if there is no linear separation at all. This will be achieved by so called *kernels*. The original space is warped into a structure which allows for linear separation. Instead of applying a support vector machine directly to the raw data $\mathbf{x}_i$, called *attributes*, it is applied to transformed data, called *features* $\Phi(\mathbf{x}_i)$. The mapping Φ is called *feature mapping*. The aim of such transformation is to achieve linear separability of the features. Two typical examples are illustrated in Figs. 6.11 and 6.12.

Some technical details must be taken into account when developing appropriate feature mappings. Let us begin by contemplating the dual problem (6.22).

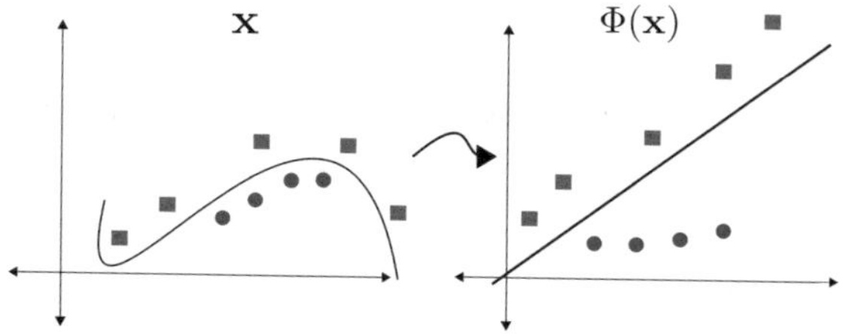

Fig. 6.11 Warping the original space by a function Φ leads to a linearly separable situation

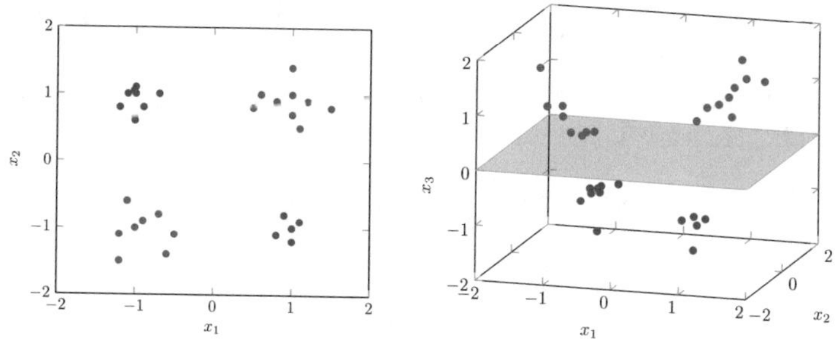

Fig. 6.12 Adding a dimension by kernel methods makes nonseparable patterns (left) linearly separable (right). Instead of considering coordinates (x_1, x_2) use (x_1, x_2, x_1x_2)

$$\max_{\boldsymbol{\lambda}} \left\{ g(\boldsymbol{\lambda}) = \sum_{i=1}^{n} \lambda_i - \frac{1}{2} \sum_{i,j=1}^{n} y_i y_j \lambda_i \lambda_j \mathbf{x}_i^T \mathbf{x}_j \right\}$$

$$\text{such that } 0 \leq \lambda_i \leq c, \quad i = 1, \ldots, n$$

$$\sum_{i=1}^{n} \lambda_i y_i = 0$$

Note that $g(\boldsymbol{\lambda})$ only depends on the inner products $\mathbf{x}_i^T \mathbf{x}_j$. The trick is to substitute $\mathbf{x}_i$ by $\Phi(\mathbf{x}_i)$ and use some inner product $\langle \cdot, \cdot \rangle$ in the corresponding image space. Hence, $\mathbf{x}_i^T \mathbf{x}_j$ is replaced by

$$K(\mathbf{x}_i, \mathbf{x}_j) = \langle \Phi(\mathbf{x}_i), \Phi(\mathbf{x}_j) \rangle.$$

The kernel $K(\mathbf{x}_i, \mathbf{x}_j)$ is often much easier to compute than $\Phi(\mathbf{x}_i)$ itself. It is mostly not necessary to explicitly represent the features in a high dimensional space. Only the values $K(\mathbf{x}_i, \mathbf{x}_j)$ must be computed.

The intuition behind the inner product approach is as follows. If $\Phi(\mathbf{x})$ and $\Phi(\mathbf{y})$ are close in the corresponding Hilbert space, then the inner product $\langle \Phi(\mathbf{x}), \Phi(\mathbf{y}) \rangle$ is large. Both $\Phi(\mathbf{x})$ and $\Phi(\mathbf{y})$ point in the same direction. If $\Phi(\mathbf{x})$ is orthogonal to $\Phi(\mathbf{y})$, then $\langle \Phi(\mathbf{x}), \Phi(\mathbf{y}) \rangle = 0$. Hence, $K(\mathbf{x}, \mathbf{y})$ measures how similar $\mathbf{x}$ and $\mathbf{y}$ are. Identifying an appropriate feature space

$$\left\{\Phi(\mathbf{x}) \mid \mathbf{x} \in \mathbb{R}^p\right\}$$

with inner product $\langle \cdot, \cdot \rangle$ is essential for successfully applying kernel transformations.

Example 6.2 Consider the feature mapping

$$\Phi(\mathbf{x}) = (x_1, x_2, x_1 x_2), \;\; \mathbf{x} = (x_1, x_2) \in \mathbb{R}^2.$$

The corresponding kernel is given by

$$K(\mathbf{x}, \mathbf{y}) = x_1 y_1 + x_2 y_2 + x_1 y_1 x_2 y_2, \mathbf{x} = (x_1, x_2), \mathbf{y} = (y_1, y_2) \in \mathbb{R}^2.$$

This example corresponds to Fig. 6.12, where linear separability in two dimensions is impossible but clearly can be achieved in three dimensions.

Example 6.3 Let $\mathbf{x} = (x_1, \ldots, x_p), \;\; \mathbf{y} = (y_1, \ldots, y_p) \in \mathbb{R}^p$. Define the kernel

$$K(\mathbf{x}, \mathbf{y}) = \Big(\sum_{i=1}^{p} x_i y_i\Big)^2.$$

The question arises whether there is some feature mapping Φ such that $K(\mathbf{x}, \mathbf{y})$ has a representation as inner product $\langle \Phi(\mathbf{x}), \Phi(\mathbf{y}) \rangle$ in some feature space. Let us investigate this question for $p = 2$ with $\mathbf{x} = (x_1, x_2)$ and $\mathbf{y} = (y_1, y_2)$. If we use

$$\Phi(\mathbf{x}) = (x_1^2, x_2^2, x_1 x_2, x_2 x_1) : \mathbb{R}^2 \to \mathbb{R}^4$$

and the Euclidean inner product $\langle \mathbf{u}, \mathbf{v} \rangle = \sum_{i=1}^4 u_i v_i$ in $\mathbb{R}^4$ then we obtain

$$\begin{aligned}
\big\langle \Phi(\mathbf{x}), \Phi(\mathbf{y}) \big\rangle &= x_1^2 y_1^2 + x_2^2 y_2^2 + x_1 x_2 y_1 y_2 + x_2 x_1 y_2 y_1 \\
&= (x_1 y_1 + x_2 y_2)^2 \\
&= \langle \mathbf{x}, \mathbf{y} \rangle^2 = K(\mathbf{x}, \mathbf{y}).
\end{aligned}$$

Hence there is an appropriate feature space which allows the kernel to be represented as some inner product.

Example 6.4 The *Gaussian kernel* is defined as

$$K(\mathbf{x}, \mathbf{y}) = \exp\Big(- \frac{\|\mathbf{x} - \mathbf{y}\|^2}{2\sigma^2}\Big),$$

$\mathbf{x}, \mathbf{y} \in \mathbb{R}^p$ and $\sigma > 0$ a parameter.

Again there is the question if there exists a feature mapping $\boldsymbol{\Phi}$ and an inner product so that the Gaussian kernel can be represented accordingly. This question will now be answered in general.

Definition 6.5 A kernel $K(\mathbf{x}, \mathbf{y})$ is called *valid* if there exists some feature function Φ such that

$$K(\mathbf{x}, \mathbf{y}) = \langle \Phi(\mathbf{x}), \Phi(\mathbf{y}) \rangle \text{ for all } \mathbf{x}, \mathbf{y} \in \mathbb{R}^p.$$

The following theorem characterizes valid kernels by the nonnegativity of a certain corresponding matrix. The original paper by Mercer [29] considers the more general situation of inner products in a function space.

Theorem 6.6 (J. Mercer)
Given $K : \mathbb{R}^p \times \mathbb{R}^p \to \mathbb{R}$. *K is a valid kernel if and only if for any* $n \in \mathbb{N}$ *and* $\mathbf{x}_1, \ldots, \mathbf{x}_n \in \mathbb{R}^p$ *the kernel matrix*

$$\mathbf{K} = \big(K(\mathbf{x}_i, \mathbf{x}_j)\big)_{i,j=1,\ldots,n}$$

is nonnegative definite.

Proof We only prove the "easy" implication from left to right. Assume that K is valid. Then there exists some feature mapping Φ such that

$$K(\mathbf{x}_i, \mathbf{x}_j) = \langle \Phi(\mathbf{x}_i), \Phi(\mathbf{x}_j) \rangle = \langle \Phi(\mathbf{x}_j), \Phi(\mathbf{x}_i) \rangle$$

by definition of the inner product. Hence, symmetry of the kernel matrix follows. Nonnegativity follows from considering the quadratic form

$$\begin{aligned} \mathbf{z}^T \mathbf{K} \mathbf{z} &= \sum_{k,\ell=1}^{n} z_k z_\ell \langle \Phi(\mathbf{x}_k), \Phi(\mathbf{x}_l) \rangle \\ &= \Big\langle \sum_{k=1}^{n} z_k \Phi(\mathbf{x}_k), \sum_{\ell=1}^{n} z_\ell \Phi(\mathbf{x}_\ell) \Big\rangle \geq 0 \end{aligned}$$

for any $\mathbf{z} = (z_1, \ldots, z_n) \in \mathbb{R}^n$, which shows the implication. Proving the reverse direction requires much deeper mathematical tools, which are not provided in the present book. ■

Example 6.7 *Polynomial kernels* are defined by

$$K(\mathbf{x}, \mathbf{y}) = (\mathbf{x}^T \mathbf{y} + c)^d, \ \mathbf{x}, \mathbf{y} \in \mathbb{R}^p, \ c \in \mathbb{R}, \ d \in \mathbb{N}.$$

The feature space has dimension $\binom{p+d}{d}$, containing all monomials of degree $\leq d$.

To determine the right kernel for a given data set is often difficult and relies on experience of the user. Sometimes it is successful to apply different kernel types to the data and observe which one provides the best separability. This is particularly true if there is no prior knowledge about the structure of the data. In this sense, finding the best kernel is part of the training process of support vector machines.

6.8 Exercises

Exercise 6.1 Let the data consist of only two points, $(\mathbf{x}_1, y_1 = +1)$ and $(\mathbf{x}_2, y_2 = -1)$. Find the SVM classifier and its parameters.

Exercise 6.2 Consider data consisting of three points only $(\mathbf{x}_i, y_i = +1)$, $i = 1, 2$ and $(\mathbf{x}_3, y_3 = -1)$. Suppose that the points are linearly separable.

(a) Show that if the points form an obtuse triangle, the maximum margin of the SVM classifier is obtained by the minimum of $\|\mathbf{x}_1 - \mathbf{x}_3\|$ and $\|\mathbf{x}_2 - \mathbf{x}_3\|$.
(b) Discuss the case where the points form an acute triangle and argue why the margin cannot be the same as in (a).

Exercise 6.3 (*Dual problems for linear and quadratic programming*)
For vectors "$\leq$" is meant componentwise.

(a) Consider the following linear programming problem:

$$\begin{aligned} \min \quad & \mathbf{c}^T\mathbf{x} \\ \text{s.t.} \quad & \mathbf{A}\mathbf{x} \leq \mathbf{b}\,. \end{aligned}$$

Determine the dual problem.
(b) Suppose that $\mathbf{B}$ is a positive definite matrix and consider the following quadratic programming:

$$\begin{aligned} \min \quad & \mathbf{x}^T\mathbf{B}\mathbf{x} \\ \text{s.t.} \quad & \mathbf{A}\mathbf{x} \leq \mathbf{b}\,. \end{aligned}$$

Determine the dual problem.
(c) For $p \geq 1$ consider the following norm minimization problem:

$$\begin{aligned} \min \quad & \|\mathbf{x}\|_p \\ \text{s.t.} \quad & \mathbf{A}\mathbf{x} = \mathbf{b}. \end{aligned}$$

Determine the dual problem.

Exercise 6.4 Consider the SVM optimization problem (6.21) for a linearly non-separable dataset:

$$\begin{aligned} \min_{\mathbf{a},b,\xi}, \quad & \frac{1}{2}\|\mathbf{a}\|^2 + c\sum_{i=1}^{n}\xi_i \\ \text{s.t.} \quad & y_i(\mathbf{a}^T\mathbf{x}_i + b) \geq 1 - \xi_i, \quad i = 1, \ldots, n \\ & \xi_i \geq 0 \quad i = 1, \ldots, n\,. \end{aligned}$$

(a) Determine the dual problem of this optimization problem.

(b) Suppose that support vectors and the optimal $\mathbf{a}^*$ are given. Find the optimal b^*.

Exercise 6.5 In each step, the SMO algorithm needs the solution of quadratic equation (6.29).

(a) Determine the coefficients $\alpha_\ell,\ \ell = 0, 1, 2,$ of Eq. (6.29) and compute the extremum.
(b) Verify the box constraints (6.31) and (6.32).

Exercise 6.6 Let $K_1 : \mathbb{R}^p \times \mathbb{R}^p \to \mathbb{R}$ and $K_2 : \mathbb{R}^p \times \mathbb{R}^p \to \mathbb{R}$ be valid kernels for a support vector machine. Show that

(a) $K(\mathbf{x}, \mathbf{y}) = \alpha K_1(\mathbf{x}, \mathbf{y})$, where $\alpha > 0$, is also a valid kernel.
(b) $K(\mathbf{x}, \mathbf{y}) = K_1(\mathbf{x}, \mathbf{y}) + K_2(\mathbf{x}, \mathbf{y})$ is also a valid kernel.
(c) $K(\mathbf{x}, \mathbf{y}) = K_1(\mathbf{x}, \mathbf{y})K_2(\mathbf{x}, \mathbf{y})$ is also a valid kernel.

Exercise 6.7 Is the Gaussian kernel in Example 6.4 valid? What feature mapping can be used? Is it unique?

Exercise 6.8 Determine the feature mapping $\Phi(\mathbf{x})$ corresponding to polynomial kernels in Example 6.7.

Chapter 7
Machine Learning

The rate of publications on machine learning has significantly increased over the last few years. Recent comprehensive books on the material are [32, 38, 41].

In this chapter we will concentrate on a few aspects which we deem important for a lecture on data science and machine learning. We will also focus on aspects which are methodologically related to data analytics and stochastic modeling of the learning process.

There are mainly two principles of machine learning: *supervised* versus *unsupervised learning*. In the first case, models depending on a potentially huge amount of parameters are trained on the basis of structurally known data so that the learned parameter set allows for generalizations on unknown test data. A model can be one of different things. For example, an artificial neural network learns the optimum weights by being told class correspondence by an expert in the training phase. Support Vector Machines learn from training data so that an optimal separating hyperplane is determined.

In unsupervised learning there is no training data with known output, predefined by an expert. However, there is feedback from the environment in the sense that a reward is given for appropriate decisions and maybe a penalty in the case that inappropriate decisions are made. What is perceived by the system is normally the actual state, and on the basis of this and past experience a decision has to made about the next action. This briefly summarize the principles of *reinforcement learning*, a typical class of unsupervised learning paradigms.

7.1 Supervised Learning

We first introduce a general model. The set $\{(\mathbf{x}_i, \mathbf{y}_i) \mid i = 1, \ldots, n\}$ is called training set, its elements $(\mathbf{x}_i, \mathbf{y}_i)$ are called *training examples* or *samples*. $\mathbf{x}_i \in \mathcal{X}$ are the *input* or *feature variables*, $\mathbf{y}_i \in \mathcal{Y}$ are the *output* or *target variables*.

R. Mathar et al., *Fundamentals of Data Analytics*,
https://doi.org/10.1007/978-3-030-56831-3_7

Supervised learning can be formalized as to determine a function

$$h : \mathcal{X} \to \mathcal{Y}$$

so that $h(\mathbf{x}_i)$ is a good predictor of $\mathbf{y}_i$. Function h is also called a *hypothesis*.

If $\mathcal{Y}$ is continuous we have a *regression* problem, if $\mathcal{Y}$ is discrete a *classification* problem evolves. In the next chapter we deal with a special but important sub-class of regression problems.

7.1.1 Linear Regression

The training examples are now $\mathbf{x}_i \in \mathbb{R}^p$ and $y_i \in \mathbb{R},\ i = 1, \ldots, n$. The hypothesis is

$$y_i = \vartheta_0 + x_{i1}\vartheta_1 + \cdots + x_{ip}\vartheta_p + \varepsilon_i = (1, \mathbf{x}_i^T)\boldsymbol{\vartheta} + \varepsilon_i,\ i = 1, \ldots, n \qquad (7.1)$$

with $\mathbf{x}_i^T = (x_{i1}, \ldots, x_{ip})$. Each output y_i is subject to some random error ε_i. $\boldsymbol{\vartheta} = (\vartheta_0, \ldots, \vartheta_p)^T$ is a parameter which has to be learned through the hypothesis $h_{\vartheta}(\mathbf{x}) = (1, \mathbf{x}^T)\boldsymbol{\vartheta}$, which is a parametric learning problem.

In order to achieve a compact representation of the problem we introduce the notation

$$\mathbf{X} = \begin{pmatrix} 1 & \mathbf{x}_1^T \\ \vdots & \vdots \\ 1 & \mathbf{x}_n^T \end{pmatrix} \in \mathbb{R}^{n\times(p+1)},\ \mathbf{y} = (y_1, \ldots, y_n)^T,\ \boldsymbol{\varepsilon} = (\varepsilon_1, \ldots, \varepsilon_n)^T \in \mathbb{R}^n.$$

The system of equations (7.1) can be written in matrix form as

$$\mathbf{y} = \mathbf{X}\boldsymbol{\vartheta} + \boldsymbol{\varepsilon}.$$

The problem is to find the best fitting $\boldsymbol{\vartheta}$, which we formalize as a least squares problem

$$\min_{\boldsymbol{\vartheta}\in\mathbb{R}^{p+1}} \|\mathbf{y} - \mathbf{X}\boldsymbol{\vartheta}\|. \qquad (7.2)$$

We determine a solution by using projections and orthogonality along two steps

(i) project $\mathbf{y}$ onto $\mathrm{Img}(\mathbf{X})$ to obtain $\hat{\mathbf{y}}$,
(ii) find some $\boldsymbol{\vartheta}$ such that $\hat{\mathbf{y}} = \mathbf{X}\boldsymbol{\vartheta}$.

To solve (i), first observe that $\mathbf{X}(\mathbf{X}^T\mathbf{X})^{-1}\mathbf{X}^T \in \mathbb{R}^{n\times n}$ is an orthogonal projection matrix onto $\mathrm{Img}(\mathbf{X})$, provided the inverse $(\mathbf{X}^T\mathbf{X})^{-1}$ exists. Hence,

$$\hat{\mathbf{y}} = \mathbf{X}(\mathbf{X}^T\mathbf{X})^{-1}\mathbf{X}^T\mathbf{y} = \arg\min_{\mathbf{z}\in\mathrm{Img}(\mathbf{X})} \|\mathbf{y} - \mathbf{z}\|.$$

To address (ii), define

$$\boldsymbol{\vartheta}^* = (\mathbf{X}^T\mathbf{X})^{-1}\mathbf{X}^T\mathbf{y}. \tag{7.3}$$

Then $\mathbf{X}\boldsymbol{\vartheta}^* = \mathbf{X}(\mathbf{X}^T\mathbf{X})^{-1}\mathbf{X}^T\mathbf{y} = \hat{\mathbf{y}}$ so that ϑ^* is an optimal solution of (7.2). The system of equations (7.3) is called *normal equations*.

Note that for the above to work the inverse $(\mathbf{X}^T\mathbf{X})^{-1}$ must exists. If this is not the case $(\mathbf{X}^T\mathbf{X})^{-1}$ can be replaces by the so called *Moore-Penrose inverse* $(\mathbf{X}^T\mathbf{X})^+$, see [30, p. 422]. The same arguments apply concerning projections so that in general $\vartheta^* = (\mathbf{X}^T\mathbf{X})^+\mathbf{X}^T\mathbf{y}$ is a solution of regression problem (7.2).

Example 7.1 (*One-dimensional linear regression*)
With $p = 1$ only two parameters ϑ_0, ϑ_1 are involved. The hypothesis is

$$y_i = \vartheta_0 + \vartheta_1 x_i + \varepsilon_i, \; i = 1, \ldots, n.$$

Further note that

$$\mathbf{X} = \begin{pmatrix} 1 & x_1 \\ \vdots & \vdots \\ 1 & x_n \end{pmatrix} \in \mathbb{R}^{n\times 2} \text{ and } \boldsymbol{\vartheta} = (\vartheta_0, \vartheta_1)^T.$$

After some algebra we obtain

$$(\mathbf{X}^T\mathbf{X})^{-1} = \frac{1}{n\sum_{i=1}^n x_i^2 - \left(\sum_{i=1}^n x_i\right)^2} \begin{pmatrix} \sum_{i=1}^n x_i^2 & -\sum_{i=1}^n x_i \\ -\sum_{i=1}^n x_i & n \end{pmatrix}. \tag{7.4}$$

With the notation

$$\bar{x} = \frac{1}{n}\sum_{i=1}^n x_i, \quad \bar{y} = \frac{1}{n}\sum_{i=1}^n y_i,$$

$$\sigma_x^2 = \frac{1}{n}\sum_{i=1}^n x_i^2 - \bar{x}^2, \quad \sigma_{xy} = \frac{1}{n}\sum_{i=1}^n x_i y_i - \bar{x}\bar{y},$$

Note that σ_x^2 is the ML estimator of the variance and σ_{xy} estimates the covariance between x- and y-values.

It finally follows from the equation $\boldsymbol{\vartheta}^* = (\mathbf{X}^T\mathbf{X})^{-1}\mathbf{X}^T\mathbf{y}$ that

$$\vartheta_1^* = \frac{\sigma_{xy}}{\sigma_x^2} \text{ and } \vartheta_0^* = \bar{y} - \vartheta_1^*\bar{x}.$$

The graph of the function $y = f(x) = \vartheta_0^* + \vartheta_1^* x$ is called the *regression line*. It predicts how the response variable y changes with an explanatory variable x. Some future observation x entails the estimated response $y = f(x)$.

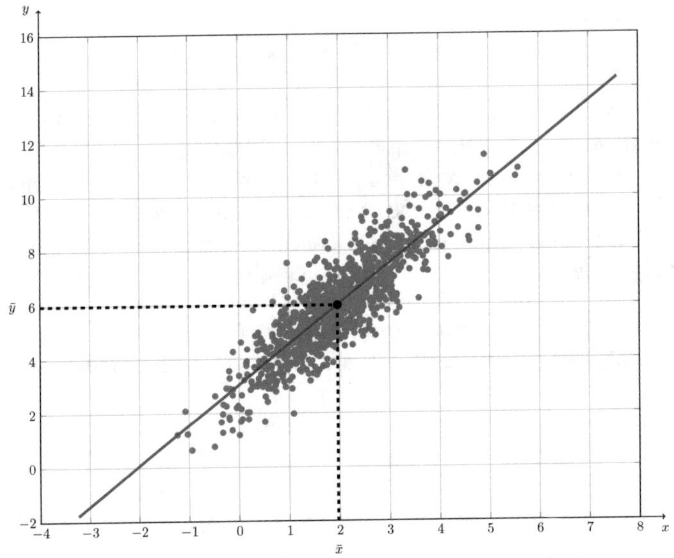

Fig. 7.1 A typical linear regression example with the regression line in red. The dotted lines indicate the two-dimensional average $(\bar{x}, \bar{y})$

Parameter ϑ_1^* determines the slope of the linear regression function, and the function graph cuts the y-axis at ϑ_0^*. Obviously $f(\bar{x}) = \bar{y}$, so that the two-dimensional center point $(\bar{x}, \bar{y})$ is part of the regression line. A typical example of a two-dimensional linear regression is shown in Fig. 7.1.

7.1.2 Logistic Regression

The final goal of *logistic regression* is binary classification between two classes 0 and 1, say. The hypothesis $h_{\boldsymbol{\vartheta}}$ leading to this outcome is again parametric. The *logistic* or *sigmoid function*

$$g(\mathbf{z}) = \frac{1}{1 + e^{-z}}, \quad z \in \mathbb{R} \tag{7.5}$$

is used to define the hypothesis

$$h_{\boldsymbol{\vartheta}}(\mathbf{x}) = g(\boldsymbol{\vartheta}^T \mathbf{x}) = \frac{1}{1 + e^{-\vartheta^T \mathbf{x}}}, \quad \mathbf{x} \in \mathbb{R}^p. \tag{7.6}$$

The image of $g(z)$ is the open interval $(0, 1)$. The function graph is depicted in Fig. 7.2. A useful feature is the form of the derivative of the logistic function, namely

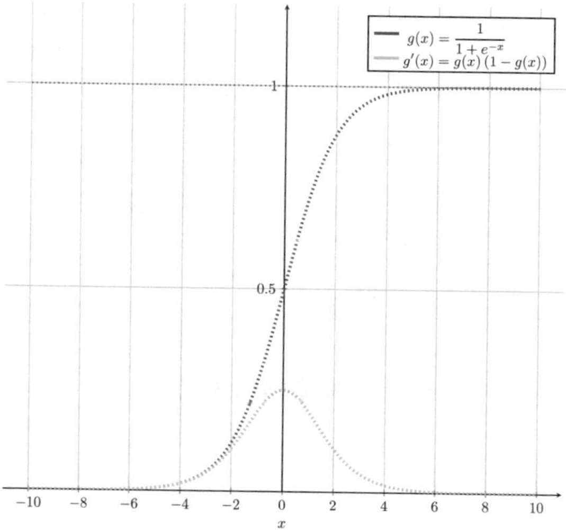

Fig. 7.2 The logistic function $g : \mathbb{R} \to (0, 1)$, $g(z) = \frac{1}{1+e^{-z}}$ and its derivative $g'(z) = g(z)\big(1 - g(z)\big)$

$$g'(z) = g(z)\,\big(1 - g(z)\big) = \frac{e^{-z}}{\left(1 + e^{-z}\right)^2}. \tag{7.7}$$

Our objective is to determine the optimal parameter ϑ for hypothesis $h_\vartheta(\mathbf{x})$ in (7.6) by some training set

$$(\mathbf{x}_i, y_i),\ \mathbf{x}_i \in \mathbb{R}^p,\ y_i \in \{0, 1\},\ i = 1, \ldots, n\,.$$

As in the previous Sect. 7.1.1 we include the constant $x_{i0} = 1$ and denote $\mathbf{x}_i = (1, x_{i1}, \ldots, x_{ip})^T$, further $\boldsymbol{\vartheta} = (\vartheta_0, \vartheta_1, \ldots, \vartheta_p)$ such that

$$\boldsymbol{\vartheta}^T \mathbf{x}_i = \vartheta_0 + \sum_{j=1}^{p} \vartheta_j x_{ij}.$$

The objective is to find the optimal match for $\boldsymbol{\vartheta}$ based on the training data. The following probabilistic interpretation allows for a precise formulation of the optimum. For a single observation $\mathbf{x}$ class correspondence $y \in \{0, 1\}$ is interpreted as a random variable, and the value of the hypothesis as probability depending on parameter $\boldsymbol{\vartheta}$ and $\mathbf{x}$.

$$\begin{aligned} P(y = 1 \mid \mathbf{x}, \boldsymbol{\vartheta}) &= h_{\boldsymbol{\vartheta}}(\mathbf{x}) \\ P(y = 0 \mid \mathbf{x}, \boldsymbol{\vartheta}) &= 1 - h_{\boldsymbol{\vartheta}}(\mathbf{x}) \end{aligned} \tag{7.8}$$

Hence, the discrete density of r.v. y given $\mathbf{x}$ and $\boldsymbol{\vartheta}$ is

$$p(y \mid \mathbf{x}, \boldsymbol{\vartheta}) = \big(h_{\boldsymbol{\vartheta}}(\mathbf{x})\big)^{y}\big(1 - h_{\boldsymbol{\vartheta}}(\mathbf{x})\big)^{(1-y)}, \ \ y = 0, 1.$$

We further assume that the training samples are independent and define

$$\mathbf{X} = \begin{pmatrix} 1 & \mathbf{x}_1^T \\ \vdots & \vdots \\ 1 & \mathbf{x}_n^T \end{pmatrix} = \big(x_{ij}\big)_{1 \le i \le n, 0 \le j \le p}.$$

Then the likelihood function has the form

$$\begin{aligned} L(\boldsymbol{\vartheta}) &= p(\mathbf{y} \mid \mathbf{X}, \boldsymbol{\vartheta}) = \prod_{i=1}^{n} p(y_i \mid \mathbf{x}_i, \boldsymbol{\vartheta}) \\ &= \prod_{i=1}^{n} \big(h_{\boldsymbol{\vartheta}}(\mathbf{x}_i)\big)^{y_i}\big(1 - h_{\boldsymbol{\vartheta}}(\mathbf{x}_i)\big)^{1-y_i}. \end{aligned}$$

The log-likelihood function is

$$\ell(\boldsymbol{\vartheta}) = \log L(\boldsymbol{\vartheta}) = \sum_{i=1}^{n} y_i \log\big(h_{\boldsymbol{\vartheta}}(\mathbf{x}_i)\big) + (1 - y_i)\log\big(1 - h_{\boldsymbol{\vartheta}}(\mathbf{x}_i)\big). \qquad (7.9)$$

Our objective is to maximize $\ell(\boldsymbol{\vartheta})$ over ϑ for given $\mathbf{y}$ and $\mathbf{X}$. Gradient ascent is a powerful method to determine the maximum in (7.9). Iteration steps are defined by

$$\boldsymbol{\vartheta}^{(k+1)} = \boldsymbol{\vartheta}^{(k)} + \alpha \nabla_{\boldsymbol{\vartheta}} \ell(\boldsymbol{\vartheta}^{(k)}), \ \ k \in \mathbb{N}.$$

Step size α plays the role of a "learning parameter".

We now compute the derivative ∇_{ϑ} for each addend in (7.9). To alleviate notation index i is skipped for the moment, hence setting $(\mathbf{x}_i, y_i) = (\mathbf{x}, y)$. The j-th partial derivative is obtained by the chain rule and using (7.7).

$$\begin{aligned} &\frac{\partial}{\partial \vartheta_j}\Big(y \log\big(h_{\boldsymbol{\vartheta}}(\mathbf{x})\big) + (1-y)\log\big(1 - h_{\boldsymbol{\vartheta}}(\mathbf{x})\big)\Big) \\ &= \Big(\frac{y}{g(\boldsymbol{\vartheta}^T\mathbf{x})} - \frac{1-y}{1 - g(\boldsymbol{\vartheta}^T\mathbf{x}}\Big)\frac{\partial}{\partial \vartheta_j} g(\boldsymbol{\vartheta}^T\mathbf{x}) \\ &= \Big(\frac{y}{g(\boldsymbol{\vartheta}^T\mathbf{x})} - \frac{1-y}{1 - g(\boldsymbol{\vartheta}^T\mathbf{x})}\Big) g(\boldsymbol{\vartheta}^T\mathbf{x})\big(1 - g(\boldsymbol{\vartheta}^T\mathbf{x})\frac{\partial}{\partial \vartheta_j}\boldsymbol{\vartheta}^T\mathbf{x} \\ &= \big(y(1 - g(\boldsymbol{\vartheta}^T\mathbf{x}) - (1-y)g(\boldsymbol{\vartheta}^T\mathbf{x})\big)\, x_j \\ &= \big(y - h_{\boldsymbol{\vartheta}}(\mathbf{bx})\big)\, x_j, \ \ j = 1, \ldots, p \end{aligned}$$

with $\mathbf{x} = (1, x_1, \ldots, x_p)^T$. Reintroducing indices per sample gives the partial derivatives of the log-likelihood function, namely

$$\frac{\partial}{\partial \vartheta_j}\ell(\boldsymbol{\vartheta}) = \sum_{i=1}^{n} \big(y_i - h_{\boldsymbol{\vartheta}}(\mathbf{x}_i)\big)x_{ij}, \; j = 1, \ldots, p\,.$$

The update rule for index k is finally obtained as

$$\vartheta_j^{(k+1)} = \vartheta_j^{(k)} + \alpha \sum_{i=1}^{n} \big(y_i - h_{\boldsymbol{\vartheta}^{(k)}}(\mathbf{x}_i)\big)x_{ij}, \; j = 1, \ldots, p \tag{7.10}$$

or in vector notation

$$\boldsymbol{\vartheta}^{(k+1)} = \boldsymbol{\vartheta}^{(k)} + \alpha \sum_{i=1}^{n} \big(y_i - h_{\boldsymbol{\vartheta}^{(k)}}(\mathbf{x}_i)\big)\mathbf{x}_i\,.$$

Recall the probabilistic interpretation $h_{\boldsymbol{\vartheta}}(\mathbf{x}) = P(y = 1 \mid \mathbf{x}, \boldsymbol{\vartheta})$, so that for a single observation $\mathbf{x}$ with class label y the steepest ascent is achieved in the direction $\mathbf{x}$ with step size $\alpha\big(y - P(y = 1 \mid \mathbf{x}, \boldsymbol{\vartheta})\big)$.

The iterative method can be improved by using Newton's method with second order derivatives. Let $\mathbf{H}$ denote the Hessian, the matrix of second order derivatives $\frac{\partial}{\partial \vartheta_i\, \partial \vartheta_j}$, $i, j = 1, \ldots, p$. Newton's update rule yields

$$\vartheta_j^{(k+1)} = \vartheta_j^{(k)} - \mathbf{H}^{(k)^{-1}} \nabla_{\boldsymbol{\vartheta}} \ell(\boldsymbol{\vartheta}^{(k)}).$$

Now let $\boldsymbol{\vartheta}^*$ denote the optimal solution, i.e.,

$$\boldsymbol{\vartheta}^* = \arg\max_{\boldsymbol{\vartheta} \in \mathbb{R}^{p+1}} \ell(\vartheta).$$

and $\mathbf{x}$ a future observation whose class correspondence shall be determined. From the probabilistic interpretation (7.8)

$$y = h_{\boldsymbol{\vartheta}^*}(\mathbf{x}) = \frac{1}{1 + e^{-\vartheta^{*T}\mathbf{x}}}$$

follows immediately as a *soft decision rule*. It means the probability that observation $\mathbf{x}$ belongs to class "1". By quantization with threshold $1/2$ the soft decision rule can be easily converted into a *hard decision* about class correspondence, as will be considered in the next section.

7.1.3 The Perceptron Learning Algorithm

In the logistic regression approach, the soft decision rule returns a value in the interval $(0, 1)$, roughly meaning the probability of class correspondence. A hard decision with values strictly 0 or 1 is enforced by setting

$$g(z) = \begin{cases} 1, & \text{if } z \geq 0 \\ 0, & \text{if } z < 0 \end{cases} .$$

In the framework of artificial neural networks, $g(z)$ is called *activation function*. Using the hypothesis

$$h_{\boldsymbol{\vartheta}}(\mathbf{x}) = g(\boldsymbol{\vartheta}^T \mathbf{x}), \ \mathbf{x} \in \mathbb{R}^p \tag{7.11}$$

and applying the same update rule as in (7.10) with h from (7.11) yields

$$\boldsymbol{\vartheta}^{(k+1)} = \boldsymbol{\vartheta}^{(k)} + \alpha \sum_{i=1}^{n} \big(y_i - h_{\boldsymbol{\vartheta}^{(k)}}(\mathbf{x}_i)\big)\mathbf{x}_i.$$

This is the so called *perceptron algorithm.* It is a rough model for how a neuron in the brain functions. Obviously it is lacking the elegant probabilistic interpretation as maximum likelihood estimator with the steepest ascent algorithm for finding an optimum $\boldsymbol{\vartheta}$. Forcing values to the boundaries 0 or 1 is interpreted for a neuron to "fire" (1) or "stay inactive" (0).

The algorithm is often applied stepwise. Parameter $\boldsymbol{\vartheta}$ is updated after having seen sample by sample $\mathbf{x}_i$ with class label y_i. This is iterated over the whole training set $i = 1, \ldots, n$. Updating $\boldsymbol{\vartheta}$ then follows the rule

$$\boldsymbol{\vartheta}^{(k+1)} = \boldsymbol{\vartheta}^{(k)} + \alpha\big(y_k - h_{\boldsymbol{\vartheta}^{(k)}}(\mathbf{x}_k)\big)\mathbf{x}_k. \tag{7.12}$$

Other choices for the activation function g are the *rectified linear* and *sigmoid function* (7.5).

7.1.4 Training a Single Neuron Binary Classifier

Artificial neural networks consist of neurons which are organized layer-wise. Each neuron receives input x_j from all predecessor neurons, forms a weighted sum with weights ϑ_j and produces output y by applying some activation function g to the weighted sum.

$$y = g\Big(\sum_{j=0}^{p} \vartheta_j x_j\Big) = g\big(\boldsymbol{\vartheta}^T \mathbf{x}\big)$$

where $\mathbf{x}$ and $\boldsymbol{\vartheta}$ are the obvious vector notations used in the previous sections. Common activation functions are the sigmoid (7.5) and the binary (7.11).

We now want to train a single neuron, i.e., find the optimal weights on the basis of some training set $(\mathbf{x}_i, y_i)$, $i = 1, \ldots, n$, see Fig. 7.3. Let us assume that $\mathbf{x}_i \in \mathbb{R}^p$ and

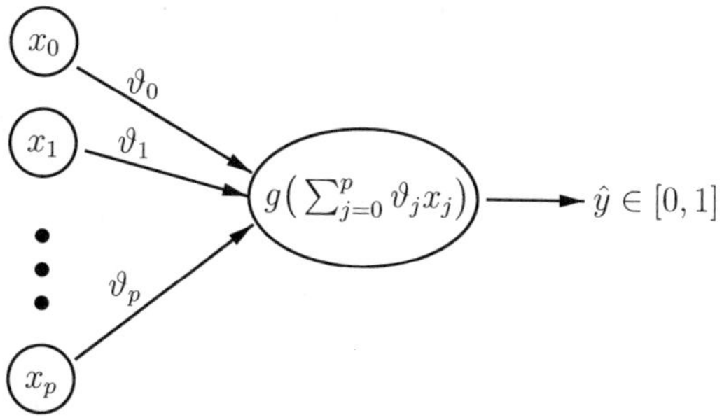

Fig. 7.3 A single neuron of an artificial neural network. x_j is the input, $\hat{y}$ the estimated class label. ϑ is optimized by training data

$y_i \in \{0, 1\}$ denotes a certain binary class allocation or label. A reasonable criterion for optimizing the weight parameter $\boldsymbol{\vartheta}$ is to

$$\max_{\boldsymbol{\vartheta} \in \mathbb{R}^p} P(y \mid \mathbf{x}, \boldsymbol{\vartheta}),$$

precisely the situation described in the previous sections. Hence, algorithm (7.12) with either the logistic or the binary threshold *activation function* can be used to train the neuron. Let $\hat{\vartheta}$ denote the vector of weights obtained after training. A new observation $\mathbf{x}$ is then soft or hard classified as

$$\hat{y} = h_{\hat{\vartheta}}(\mathbf{x}) = g(\hat{\boldsymbol{\vartheta}}^T \mathbf{x})$$

depending on the employed activation function.

7.2 Reinforcement Learning

Reinforcement learning does not need a-priori training data. Instead a scheme is needed by which actions of the learning system are rewarded or penalized. Think of how a baby learns to walk. Toppling over is a negative reward for uncontrolled balance, walking over a certain distance is the reinforcement for further correct action.

The present state is known to the system. It then chooses some action to reach the next state according to a certain probability distribution, which depends on the present state and the chosen action. Each state and action is connected to a reward and the goal for the system is to maximize the total expected reward.

It is quite natural to include randomness into the model because reaching the successor state according to an action is also influenced by the environment and cannot be precisely predicted. Consider, e.g., learning coordination of robot legs or arms, autonomous flying or even factory control. The appropriate mathematical model for including these effects are Markov decision processes.

7.2.1 Markov Decision Processes

A *Markov decision process* (MDP) is described by a 5-tuple

$$\left(\mathcal{S}, \mathcal{A}, P_{s,a}, \gamma, R\right) \tag{7.13}$$

with the following interpretation:

$\mathcal{S}$	finite set of states
$\mathcal{A}$	finite set of actions
$P_{s,a}(t),\ s, t \in \mathcal{S},\ a \in \mathcal{A}$	transition probabilities
$\gamma \in [0, 1]$	discount factor
$R : \mathcal{S} \times \mathcal{A} \to \mathbb{R}$	reward function

In each of the states $s \in \mathcal{S}$ an action $a \in \mathcal{A}$ can be chosen. The distribution $P_{s,a}(t),\ t \in \mathcal{S}$, contains the conditional probabilities to transition from state s to state t when taking action a. The reward function R determines the reward which is received in state s and taking action a. The purpose of the discount factor γ is to mitigate the influence of past rewards. This will become clearer if we discuss the total payoff. In many cases the reward function depends only on the state such that $R : \mathcal{S} \to \mathbb{R}$ has preimage $\mathcal{S}$ only.

The dynamics of the MDP can be visualized as a chain

$$s_0 \xrightarrow{a_0} s_1 \xrightarrow{a_1} s_2 \xrightarrow{a_2} s_3 \xrightarrow{a_3} \cdots$$

where $s_0, s_1, s_2, \ldots$ is the series of states attained by the system and $a_0, a_1, a_2, \ldots$ the series of actions chosen in the corresponding states.

The total *payoff* is defined as

$$R(s_0, a_0) + \gamma R(s_1, a_1) + \gamma^2 R(s_2, a_2) + \gamma^3 R(s_3, a_3) + \cdots .$$

Discount factor γ geometrically reduces the influence of past rewards. The motivation is that human behavior shows preference for immediate reward. Moreover, it avoids infinite rewards for periodic cycles of the Markov chain.

If the reward function only depends on the states and not on the actions, then the payoff becomes

$$R(s_0) + \gamma R(s_1) + \gamma^2 R(s_2) + \gamma^3 R(s_3) + \cdots .$$

In order not to overload notation we will keep to this case in the sequel.

Obviously, the total payoff is a random variable with distribution depending on the initial state and action (s_0, a_0), and the transition probabilities $P_{s,a}(t),\ t \in \mathcal{S}$.

Let $S_0, S_1, S_2 \ldots$ denote the sequence of random variables describing the states attained by the system at subsequent time steps. This sequence forms a Markov chain.

The ultimate goal is to maximize the expected payoff over all possible series of actions, in detail

$$\max_{\mathbf{a}\in\mathcal{A}^\infty} E\big[R(S_0) + \gamma R(S_1) + \gamma^2 R(S_2) + \gamma^3 R(S_3) + \cdots\big] \tag{7.14}$$

where $\mathbf{a} = (a_0, a_1, a_2, \ldots)$ denotes an infinite sequence of actions in $\mathcal{A}^\infty$.

A *policy* is a function mapping states $s \in \mathcal{S}$ onto actions $a \in \mathcal{A}$,

$$\pi : \mathcal{S} \to \mathcal{A} : s \mapsto a = \pi(s).$$

It determines which action will be taken if the system attains state s.

Based on the policy function the *value function* for a policy π is defined on the set of states as

$$V^\pi(s) = E\big[R(S_0) + \gamma R(S_1) + \gamma^2 R(S_2) + \gamma^3 R(S_3) + \cdots \mid S_0 = s, \pi\big].$$

It is the expected total payoff upon starting in s and applying policy π. For any policy π the following *Bellman equations* hold.

Theorem 7.2 (Bellman equations)

$$V^\pi(s) = R(s) + \gamma \sum_{t\in\mathcal{S}} P_{s,\pi(s)}(t)\, V^\pi(t),\;\; s \in \mathcal{S} \tag{7.15}$$

Proof We use basic properties of conditional expectations and the Markov property to obtain the following.

$$\begin{aligned}
V^\pi(s) &= R(s) + E\big[\gamma R(S_1) + \gamma^2 R(S_2) + \gamma^3 R(S_3) + \cdots \mid S_0 = s, \pi\big] \\
&= R(s) + \gamma\, E\big[R(S_1) + \gamma R(S_2) + \gamma^2 R(S_3) + \cdots \mid S_0 = s, \pi\big] \\
&= R(s) + \gamma \sum_{t\in\mathcal{S}} P_{s,\pi(s)}(t) E\big[R(S_1) + \gamma R(S_2) + \gamma^2 R(S_3) + \cdots \mid S_0 = s,\, S_1 = t, \pi\big] \\
&= R(s) + \gamma \sum_{t\in\mathcal{S}} P_{s,\pi(s)}(t) E\big[R(S_1) + \gamma R(S_2) + \gamma^2 R(S_3) + \cdots \mid S_1 = t, \pi\big] \\
&= R(s) + \gamma \sum_{t\in\mathcal{S}} P_{s,\pi(s)}(t) V^\pi(t)
\end{aligned}$$

In the third line the *law of total probability* is used. ■

The right hand side of Eq. (7.15) is the immediate reward $R(s)$ plus the expected sum of future discounted rewards.

If $\mathcal{S}$ is finite, $\mathcal{S} = \{1, 2, \ldots, m\}$, say, solving the Bellman equations for V^π with a fixed policy π means to solve $|\mathcal{S}|$ linear equations with $|\mathcal{S}|$ variables. To see this write

$$\mathbf{R} = \big(R(1), \ldots, R(m)\big)^T \in \mathbb{R}^m$$

$$\boldsymbol{\Psi}^\pi = \begin{pmatrix} P_{1,\pi(1)}(1) & \cdots & P_{1,\pi(1)}(m) \\ \vdots & \ddots & \vdots \\ P_{m,\pi(m)}(1) & \cdots & P_{m,\pi(m)}(m) \end{pmatrix} \in \mathbb{R}^{m\times m}$$

$$\mathbf{V}^\pi = \big(V^\pi(1), \ldots, V^\pi(m)\big)^T \in \mathbb{R}^m.$$

Vector $\mathbf{V}^\pi$ contains the variables to be determined. The Bellman equations (7.15) can then be written in matrix form

$$\mathbf{V}^\pi = \mathbf{R} + \gamma \boldsymbol{\Psi}^\pi \mathbf{V}^\pi \tag{7.16}$$

or equivalently

$$\mathbf{V}^\pi = \big(\mathbf{I} - \gamma \boldsymbol{\Psi}^\pi\big)^{-1} \mathbf{R} \tag{7.17}$$

provided the inverse exists.

Now, the *optimal value function* over all policies is

$$\begin{aligned} V^*(s) &= \max_\pi V^\pi(s) \\ &= \max_\pi E\big[R(S_0) + \gamma R(S_1) + \gamma^2 R(S_2) + \cdots \mid S_0 = s, \pi\big] \end{aligned} \tag{7.18}$$

representing the maximum expected payoff over all policies π when starting in state s. For the optimal value function the following *Bellman optimality equations* hold.

Theorem 7.3 (Bellman optimality equations) *The optimal value function satisfies the the fixed-point equation*

$$V^*(s) = R(s) + \max_{a\in\mathcal{A}} \Big\{\gamma \sum_{t\in\mathcal{S}} P_{s,a}(t)\, V^*(t)\Big\}. \tag{7.19}$$

Proof Denote by π' any policy which satisfies $\pi'(s) = a$ for some fixed $s \in \mathcal{S}$ and $a \in \mathcal{A}$. By the Bellman equations (7.15) it holds for any policy π that

$$\begin{aligned} V^*(s) &= \max_\pi V^\pi(s) \\ &= \max_\pi \Big\{R(s) + \gamma \sum_{t\in\mathcal{S}} P_{s,\pi(s)}(t)\, V^\pi(t)\Big\} \\ &= R(s) + \max_\pi \Big\{\gamma \sum_{t\in\mathcal{S}} P_{s,\pi(s)}(t)\, V^\pi(t)\Big\} \\ &= R(s) + \max_{a,\pi'} \Big\{\gamma \sum_{t\in\mathcal{S}} P_{s,a}(t)\, V^{\pi'}(t)\Big\} \\ &= R(s) + \max_a \Big\{\gamma \sum_{t\in\mathcal{S}} P_{s,a}(t) \max_{a,\pi'} V^{\pi'}(t)\Big\} \end{aligned}$$

$$= R(s) + \max_a \left\{ \gamma \sum_{t \in S} P_{s,a}(t)\, V^*(t) \right\}$$

This concludes the proof. ■

The proof follows Bellman's Optimality Principle: "An optimal policy has the property that, whatever the initial state and the initial decision are, the remaining decisions must constitute an optimal policy with regard to the state resulting from the first decision."

Solving the system of equations (7.19) yields the optimal value function. In contrast to the Bellman equations (7.15) the Bellman optimality equations are highly non-linear due to the maximum.

The optimal value function may hence be decomposed into the immediate reward plus the maximum over all actions a of the expected sum of discounted rewards we receive in the future upon applying action a.

We also define $\pi^*(s)$ as the policy where the maximum of the sum in (7.19) is attained, i.e.,

$$\pi^*(s) = \arg\max_{a \in \mathcal{A}} \left\{ \sum_{t \in S} P_{s,a}(t)\, V^*(t) \right\}. \tag{7.20}$$

A policy π will be preferable over another if it achieves a greater $V^{\pi}(s)$ for any initial state s. We hence define a partial ordering "$\succeq$" on the set of all policies by

$$\pi \succeq \pi' \text{ if } V^{\pi}(s) \geq V^{\pi'}(s) \text{ for all } s \in \mathcal{S}.$$

The question arises if there is a uniformly optimal policy π^* in the sense that it is preferable over all other policies. There indeed exists such optimum policy.

Theorem 7.4 (a) *There exists an optimal policy π^* such that $\pi^* \succeq \pi$ for all other policies π. The policy π^* defined in Eq. (7.20) is optimal in this sense.*
(b) *All optimal policies π^* achieve the optimal value function $V^{\pi^*}(s) = V^*(s)$, $s \in \mathcal{S}$.*

Note that policy π^* in (7.20) is optimal for all states $s \in \mathcal{S}$. This means that the same policy π^* attains the maximum in Eq. (7.18) for all states s. It is not the case that starting in some state s would lead to another optimal policy than starting in s'.

It still remains to explicitly compute an optimal policy. There are mainly two algorithms for doing so: *value iteration* and *policy iteration*. The latter will be briefly introduced next.

7.2.2 Computing an Optimal Policy

We briefly describe the principles of a powerful algorithm for computing an optimal policy π^*. Given a Markov decision process $\left(\mathcal{S}, \mathcal{A}, P_{s,a}, \gamma, R\right)$ from (7.13). The policy iteration algorithm iterates along the following steps.

1. initialize π at random
2. repeat until convergence

   (a) $\mathbf{V} := \mathbf{V}^\pi$

   (b) For each $s \in \mathcal{S}$, let $\pi(s) := \arg\max_{a\in\mathcal{A}} \sum_{t\in\mathcal{S}} P_{s,a}(t)V(t)$

If the number of states is finite, which is most relevant for practical applications, $\mathbf{V}^\pi$ can be found by solving (7.16) or (7.17). Further note that $V(t)$ is computed according to the actual policy π. The update of π is greedy w.r.t. $\mathbf{V}$. In the above algorithm, $\mathbf{V}$ converges to $\mathbf{V}^*$ and π to π^*.

7.3 Exercises

Exercise 7.1 Let $\mathbf{X}$ be a matrix in $\mathbb{R}^{p\times n}$ such that $(\mathbf{X}^T\mathbf{X})$ is invertible. Show that $\mathbf{X}(\mathbf{X}^T\mathbf{X})^{-1}\mathbf{X}^T$ is the projection matrix onto the image of $\mathbf{X}$.

Exercise 7.2 Prove the form of the inverse in (7.4). Under which conditions does the inverse matrix not exist?

Exercise 7.3 Consider the dataset which maps the average number of weekly hours ten students have spent on exercises for a certain class with the result of an exam (passed = 1, failed = 0).

No. of hours	1.2	1.7	1.1	2.2	3.4	5.0	2.6	1.3	2.8	1.1
Exam result	1	0	0	1	1	0	1	1	1	0

Use logistic regression to find an optimum decision rule.

Exercise 7.4 Consider the situation depicted in Fig. 7.4. A customer wants to park his car as close as possible to a restaurant. There are N parking lots in a row, some of them are taken by cars of other customers. The driver cannot see whether a place is available unless he is in front of it. At each place the driver can either park, if the lot is available, or move forward to next one. The closer he comes to the restaurant the higher is his reward, ranging from 1 to N. If he does not park anywhere, he leaves the place and his reward is 0.

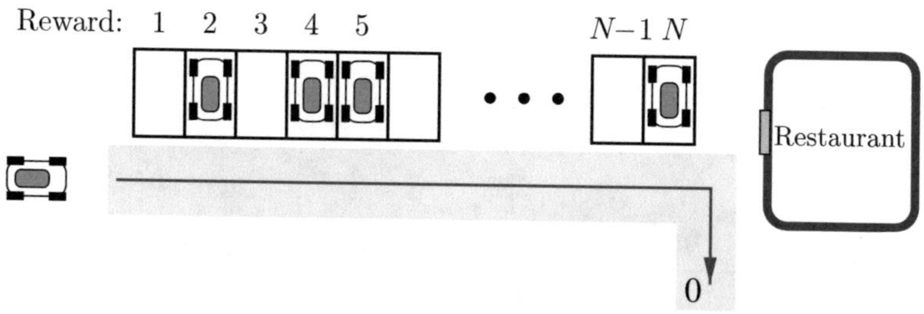

Fig. 7.4 The optimum parking problem

Find the right assumptions to model the situation as a Markov Decision Process. Can you determine an optimal policy?

References

1. Aggarwal, C.C.: Data Mining. Springer International Publishing (2015)
2. Bai, Z., Silverstein, J.W.: Spectral Analysis of Large Dimensional Random Matrices, 2nd edn. Springer, New York (2010)
3. Baik, J., Ben Arous, G., Péché, S.: Phase transition of the largest eigenvalue for nonnull complex sample covariance matrices. Ann. Probab. **33**(5), 1643–1697 (2005). https://doi.org/10.1214/009117905000000233
4. Balasubramanian, M., Schwartz, E.L.: The IsoMap algorithm and topological stability. Science **295**(5552) (2002). https://doi.org/10.1126/science.295.5552.7a
5. Bandeira, A.S.: Ten Lectures and Forty-Two Open Problems in the Mathematics of Data Science (2008). http://www.cims.nyu.edu/~bandeira/TenLecturesFortyTwoProblems.pdf
6. Birkhoff, G.: Tres observaciones sobre el algebra lineal. Univ. Nac. Tucumán, Ser. A **5**, 147–151 (1946)
7. Boser, B., Guyon, I., Vapnik, V.: A training algorithm for optimal margin classifiers. In: Proceedings of the Fifth Annual Workshop on Computational Learning Theory. Pittsburgh (1992)
8. Boyd, S., Vandenberghe, L.: Convex Optimization, 7th edn. Cambridge University Press, Cambridge (2009)
9. Cattell, R.B.: The scree test for the number of factors. Multivar. Behav. Res. **1**(2), 245–276 (1966). https://doi.org/10.1207/s15327906mbr0102_10
10. Coifman, R.R., Lafon, S.: Diffusion maps. Appl. Comput. Harmonic Anal. **21**(1), 5–30 (2006). https://doi.org/10.1016/j.acha.2006.04.006
11. Courant, R.: Über die eigenwerte bei den differentialgleichungen der mathematischen physik. Mathematische Zeitschrift **7**, 1–57 (1920). https://doi.org/10.1007/BF01199396
12. Eckart, C., Young, G.: The approximation of one matrix by another of lower rank. Psychometrika **1**(3), 211–218 (1936). https://doi.org/10.1007/BF02288367
13. Fan, K.: On a theorem of Weyl concerning eigenvalues of linear transformations I. Proc. Natl. Acad. Sci. **35**(11), 652–655 (1949). https://doi.org/10.1073/pnas.35.11.652
14. Farahat, H.K., Mirsky, L.: Permutation endomorphisms and refinement of a theorem of Birkhoff. Math. Proc. Camb. Philos. Soc. **56**(4), 322–328 (1960). https://doi.org/10.1017/S0305004100034629
15. Fischer, E.: Über quadratische formen mit reellen koeffizienten. Monatshefte für Mathematik und Physik **16**, 234–249 (1905). https://doi.org/10.1007/BF01693781
16. Gershgorin, S.A.: über die abgrenzung der eigenwerte einer matrix. Bulletin de l'Académie des Sciences de l'URSS. Classe des sciences mathématiques et na **6**, 749–754 (1931)

R. Mathar et al., *Fundamentals of Data Analytics*,
https://doi.org/10.1007/978-3-030-56831-3

17. Hardy, G.H., Littlewood, J.E., Pólya, G.: Inequalities, 1st edn. Cambridge University Press, Cambridge (1934)
18. Hastie, T., Tibshirani, R., Friedman, J.H.: The Elements of Statistical Learning: Data Mining, Inference, and Prediction. Springer Series in Statistics, 2nd edn. Springer, New York (2009)
19. Horn, R.A., Johnson, C.R.: Matrix Analysis, 2nd edn. Cambridge University Press, Cambridge (2013)
20. Kannan, R., Vempala, S.: Spectral Algorithms. now publishers, Boston, Delft (2009)
21. Leskovec, J., Rajaraman, A., Ullman, J.D.: Mining of Massive Datasets, 2nd edn. Cambridge University Press, Cambridge (2014)
22. Lloyd, S.: Least squares quantization in pcm. IEEE Trans. Inf. Theory **28**(2), 129–137 (1982)
23. Löwner, K.: Über monotone Matrixfunktionen. Mathematische Zeitschrift **38**, 177–216 (1934). http://eudml.org/doc/168495
24. Magnus, J.R., Neudecker, H.: Matrix Differential Calculus with Applications in Statistics and Econometrics, 3rd edn. Wiley, Chichester (2019)
25. Marčenko, V.A., Pastur, L.A.: Distribution of eigenvalues for some sets of random matrices. Math. USSR-Sbornik **1**(4), 457–483 (1967). https://doi.org/10.1070/sm1967v001n04abeh001994
26. Mardia, K.V., Kent, J.T., Bibby, J.M.: Multivariate Analysis. Probability and Mathematical Statistics. Academic Press, London; New York (1979)
27. Marshall, A.W., Olkin, I.: Inequalities: Theory of Majorization and Its Applications. Academic Press, New York (1979)
28. Mathar, R.: Multidimensionale Skalierung: mathematische Grundlagen und algorithmische Aspekte. Teubner-Skripten zur mathematischen Stochastik. Teubner, Stuttgart (1997). OCLC: 605887892
29. Mercer, J.: Functions of positive and negative type and their connection with the theory of integral equations. Philos. Trans. R. Soc. A **209**, 415–446 (1909)
30. Meyer, C.D.: Matrix Analysis and Applied Linear Algebra. SIAM, Philadelphia (2000)
31. Mises, R.V., Pollaczek-Geiringer, H.: Praktische verfahren der gleichungsauflösung. ZAMM - J. Appl. Math. Mech./ Zeitschrift für Angewandte Mathematik und Mechanik **9**(2), 152–164 (1929). https://doi.org/10.1002/zamm.19290090206
32. Murphy, K.P.: Machine Learning: A Probabilistic Perspective. Adaptive Computation and Machine Learning Series. MIT Press, Cambridge (2012)
33. Paul, D.: Asymptotics of sample eigenstructure for a large dimensional spiked covariance model. Statistica Sinica **17**, 1617–1642 (2007)
34. Platt, J.: Sequential minimal optimization: a fast algorithm for training support vector machines. Technical Report MSR-TR-98-14, Microsoft Research (1998)
35. Richter, H.: Zur abschützung von matrizennormen. Mathematische Nachrichten **18**(1–6), 178–187 (1958). https://doi.org/10.1002/mana.19580180121
36. Schoenberg, I.: Remarks to Maurice frechet's article "Sur la definition axiomatique d'une classe d'espace distances vectoriellement applicable sur l'espace de Hilbert". Ann. Math. **36**(3), 724–732 (1935). https://doi.org/10.2307/1968654
37. Sedgewick, R., Wayne, K.: Algorithms, 4th edn. Addison-Wesley, Upper Saddle River (2011)
38. Shalev-Shwartz, S., Ben-David, S.: Understanding Machine Learning, from Theory to Algorithms, 8th printing, 2018th edn. Cambridge University Press, Cambridge (2014)
39. Tenenbaum, J.B., De Silva, V., Langford, J.C.: A global geometric framework for nonlinear dimensionality reduction. Science **290**(5500), 2319–2323 (2000). http://science.sciencemag.org/content/290/5500/2319.short
40. Torgerson, W.S.: Multidimensional scaling: I. theory and method. Psychometrika **17**(4), 401–419 (1952). https://doi.org/10.1007/BF02288916
41. Watt, J., Borhani, R., Katsaggelos, A.: Machine Learning Refined, Foundations, Algorithms, and Applications. Cambridge University Press, Cambridge (2016)

Index

R. Mathar et al., *Fundamentals of Data Analytics*,
https://doi.org/10.1007/978-3-030-56831-3

P

R

S

T

U

V

W

Printed in the United States
by Baker & Taylor Publisher Services